The *Future* of the Semiconductor Computer Robotics and Telecommunications *Industries*

Compiled by
the Editorial Staff
of Petrocelli Books

PETROCELLI BOOKS
Princeton, New Jersey
A Source Book

The majority of the information
contained in this volume was originally published
by the U.S. Department of Commerce.

Book design by Liz Waite
Composition by Eastern Graphics

Printed in the United States of America.
iu ɤ 8 7 6 5 4 3 2 1

Library of Congress Cataloging in Publication Data
Main entry under title:

The Future of the semiconductor, computer, robotics,
    and telecommunications industries.

    1. Semiconductor industry—United States—Congresses.
2. Computer industry—United States—Congresses.
3. Robot industry—United States—Congresses.
4. Telecommunication equipment industry—United
States—Congresses. I. Petrocelli Books (Firm)
HD9696.S43U4817   1984   338.4'762138'0973 84-1085
ISBN 0-89433-259-7

Contents

CONTENTS

Part 3

The Robotics Industry

Part 4

The Telecommunications Industry

CONTENTS

Editors' Introduction

It would be difficult to identify four areas more critical to high technology than semiconductors, computers, robotics, and telecommunications. These four highly interrelated areas in many respects comprise the infrastructure of our entire technological society. Recognizing the importance of these areas, the U.S. Department of Commerce's International Trade Administration sponsored, in 1983, a number of meetings intended to shed some light on the directions in which these four areas were moving. The "high technology meetings," as they came to be known, yielded profiles of the four industries, provided a forum for discussion among government and industry leaders, and identified a set of action items for those committed to progress in the four industries.

EDITORS' INTRODUCTION

The results of the four meetings are integrated here for the first time. This book contains all of the results of the four meetings as well as some insight into how they might interrelate. They are presented here not in the order in which they occurred, but rather in an order that makes the most sense from a high technology perspective.

The first industry covered is the semiconductor industry. This section of the book contains presentations by Dr. Robert N. Noyce, Vice Chairman of the Intel Corporation, Mr. W. J. Sanders, III, Chairman of Advanced Micro Devices, Mr. Gary L. Tooker, Senior Vice President of Motorola, and Mr. Erich Bloch, Vice President of Technical Personnel of the IBM Corporation. Robert Noyce talks about the semiconductor industry in relatively general terms. After developing a short history of the semiconductor industry, he notes that there are currently a number of problems which must be solved if the industry is to endure. One such problem is the shortage of trained personnel. A second involves the lack of attention to long-term financial and industrial goals, while the third concerns export controls and technology transfer. In many respects Noyce's thoughts may be regarded as a warning about the threats to the semiconductor industry. They also lead perfectly into those of W.J. Sanders who examines in some detail the international trade policy that affects the semiconductor industry. Unlike Noyce, Sanders gets very specific and details the nature of the Japanese threat to the American semiconductor industry. He also lists no less than six recommendations for U.S. policymakers, including policies to protect U.S. industry against foreign industrial targeting practices, policies that would establish a monitoring system with respect to major commercial semiconductor product lines, policies that require that U.S. semiconductor firms receive commercial opportunities in Japan that are fully equivalent to those enjoyed by Japanese firms, policies to enforce agreements previously made by Japan regarding international competition, policies that extend our policies toward Japan to other countries, and legislation that provides the authority and means necessary to insure that the U.S. government carries out the above policies and guidelines. More specifically, Sanders suggests that the High Technology Trade Act of 1982 be reintroduced to the Congress.

EDITORS' INTRODUCTION

Gary Tooker's contribution is in the area of investment and tax policy. Like Sanders, Tooker is concerned about the differences between the Japanese and American economic systems and the advantages that accrue to the Japanese because of these differences. He is especially concerned with the role that risk plays in the two economies, citing that Japanese industrialists generally face much less risk than their American counterparts. He bluntly suggests that "the Japanese government has taken the risk out of investing in target industries in Japan." Significantly, Tooker does not suggest that the U.S. government eliminate risk from U.S. investments but rather that it become aware of the competitive environment when considering tax measures that impact high technology industries.

Erich Bloch is less concerned with taxes and risk than he is with research and education. He is concerned that a smaller and smaller percentage of resources is being allocated to research and education in the semiconductor industry. He notes that while the U.S. semiconductor industry spends approximately 10 percent of its sales dollars on research, the Japanese spend between 13 and 16 percent. Bloch suggests that the percentage be increased and that a number of unique arrangements be made with some government agencies to make certain that basic research in semiconductors continues at an aggressive pace. With regard to scientific and technical education, Bloch laments the reduction of government support to universities and suggests that because of the cutbacks new relationships between industry and academia need to be developed.

All in all, the views from the semiconductor industry expressed here are not very optimistic. It is noteworthy that they all point to a number of critical threats on the horizon, not the least of which are traceable to the East. Indeed, in many respects, the four contributions constitute a call to arms in the semiconductor industry.

Part Two of the book looks at the computer industry. Like Part One, this part begins with a profile of the industry at this point in time. This profile, prepared by Mr. Robert Eckelmann of the Office of International Sector Policy of the International Trade Administration, Department of Commerce, diagnoses the competitive position of the U.S. computer industry. It describes the industry and contrasts it with the world marketplace. It also devotes special attention to the subject of world computer trade. Eckelmann's country-by-country analysis is valuable insofar as he identifies the Japanese, French, West Germans, and British as key competitors with unique advantages and disadvantages vis-à-vis U.S. industry. All of these discussions contain statistics about the size and nature of foreign computer producers and consumers. Eckelmann concludes his study by identifying the key issues likely to affect U.S. competitiveness in international computing. These issues include the continued development and implementation of the systems concept, the shortage of trained personnel, research and development, the design and production of adequate software, and foreign targeting practices.

This excellent discussion is followed by a presentation by Vico E. Henriques who is the President of The Computer and Business Equipment Manufacturers Association (CBEMA). Among other contributions, Henriques presents a statistical tour of the computer industry, noting that industry revenues worldwide have risen dramatically since 1965 and that in all likelihood they will continue to rise just as dramatically through the 1980s and well into the 1990s. He also notes that the number of people employed in the computer industry has risen steadily since 1960.

The next contribution by Edson de Castro of the Data General Corporation looks at the restrictions and performance requirements of the computer industry. Of special interest is the identification of Brazil, Mexico, and South Korea as three of the most current examples of countries that deny or impede access to their markets.

Stephen G. Jerritts, Senior Vice President of Honeywell, develops a response to hostile industry targeting practices. He specifically looks at foreign trade barriers, fiscal policy, government-supported research, and intra-industry cooperation as four areas that can be developed in response to foreign targeting practices.

The last contribution to this section on the computer industry is made by John W. Lacey, Executive Vice President, Technology and Planning, Control Data Corporation. Lacey addresses eleven areas relevant to international trade and export controls as they impact the computer industry. The areas include the need for a positive balance of trade, the importance of multilateral controls, the importance of foreign policy controls, the need to control exports to

EDITORS' INTRODUCTION

COCOM countries, the importance of re-export controls within COCOM countries, the establishment of licensing requirements regarding exports to Third World countries, the importance of foreign market assessments, the role of indexing, the proper administration of unilateral export controls, the development of a list of critical military technologies, and the enforcement of laws and policies.

Nearly all of the contributions in this section of the book express a concern over the international position of U.S. industry. There is a great deal of apprehension about the growing capabilities of the Japanese and Western Europeans to capture more and more of the world computer market. The calls for financial and policy support frequently call for an expanded government role in the management of international trade, a role made necessary by the actions of foreign governments.

Part Three of this book turns its attention to the robotics industry. The first contribution in this section was prepared by Robert Eckelmann whose study deals with the competitive prospects of the U.S. robotics industry. He first addresses some definitional problems, i.e., what is a robot and what are the components of the robotics industry, and then develops a summary of the U.S. robotics market. This analysis is followed by a summary of the world robotics marketplace which is comprised of aggregate trends, and offers a seven-country analysis. The countries examined include Japan, West Germany, Sweden, The United Kingdom, France, Italy, and a few other relatively minor competitors. Eckelmann then looks at the world trade situation in robotics and identifies the key issues affecting U.S. competitiveness in that market. He cites the importance of international agreements between private firms, labor, and automation.

Joseph Engelberger, the well-known President of Unimation, Inc., suggests in the next contribution that the Japanese hold an impressive edge in the implementation of robotics, that the levels of technology in the U.S., Europe, and Japan are approximately even, and that the Japanese, British, German, and French governments have already launched aggressive programs to support robotics technology and to provide both internal and external marketing aids. These and other observations are examined more thoroughly by

Walter Weisel, President of Prab Robots, Inc. More specifically, Weisel suggests that the U.S. needs to strengthen its base of manufacturers, and develop a strategy for combating the aggressive posture of the Japanese vis-à-vis the exportation of robotic equipment. Stanley J. Polcyn, President of the Robotics Institute of America and Senior Vice President of Unimation, Inc., suggests that the U.S. industry/government relationship be strengthened in the following ways: the establishment of robotics as a major strategic industry; the setting of national productivity goals that will enhance the U.S. robotics industrial base; the establishment of an Industrial Policy Board (similar to the Federal Reserve Board) that could take forceful, effective and independent action to strengthen strategic industries; the development of an aggressive, consistent taxation policy that would provide incentives to utilize robotics technology; the creation and funding of a robot leasing company that would provide low cost loans to purchasers of industrial robots manufactured in the U.S.; the allocation of government funds to assist in generic, basic and applied research for advancing robotic developments; and the development of a realistic export program.

Semiconductors, computers, and robots share a great deal in common, but without telecommunications their contributions to industrial productivity are severely constrained. In fact, prudent observers frequently regard telecommunications as the high technology circulatory system. Part Four of this book concentrates on telecommunications and what the U.S. must do to maintain its edge in the industry. The section begins with a contribution by Robert Eckelmann. This opening study examines the international competitive position of the U.S. telecommunications equipment industry. It studies the world market, the nature of telecommunications trade, and the role which the U.S. plays in the world marketplace. It also focuses specifically on Japan and Western Europe, and identifies a number of key issues in the world telecommunications market, including the liberalization of services and equipment, technological change, research and development, and foreign targeting practices, among others. Eckelmann's industry profile serves as the backdrop for John Sodolski's contribution. Sodolski is the Vice President of the Telecommunications Group of the Electronic Industries Association. His contribution focuses upon the market for telecommunications equipment, and the processes by which the marketplace evolves and changes. Like so many of his counterparts in the semiconductor, computer, and robotics industries, Sodolski is very concerned about how changes in the character of the marketplace will affect the U.S. position.

Richard Moley, Vice President of Marketing of the Rolm Corporation, suggests that one way to respond to the ever-changing marketplace is to open previously closed markets. He cites at least four major obstacles to opening closed markets. They include political obstacles, where the target government decides who produces what and for how much; economic obstacles, where there are high tariffs and/or local requirements which create a very high risk for U.S. manufacturers; technical obstacles, where either specifications are unavailable to foreign suppliers or the cost of meeting unique specifications cannot be justified due to market size and potential; and technology transfer obstacles, which are being raised more and more frequently by newly industrialized countries. If the U.S. is to continue to succeed in the telecommunications marketplace, it must, according to Moley, learn how to deal with these and other obstacles. Citing a successful Rolm experience with the Japanese, he also suggests that success is by no means impossible. In fact, if anything, Moley's message is basically optimistic.

John F. Mitchel, President of Motorola, is not nearly as sanguine about the future. He bluntly states that "Motorola is doing everything it can to overcome [the] fundamental imbalance in [the] competitive situation." He also feels that existing laws are inadequate to deter foreign targeting practices and that legislation is desperately needed to develop a balance between U.S. and foreign competition. Mitchel concludes with a six-point plan about how to improve the international situation from the perspective of the U.S. telecommunications industry.

The next contribution by John N. Lemasters, Senior Vice President, Communications Sector, Harris Corporation, deals with export financing and licensing considerations in the telecommunications industry. Like many of his counterparts, Lemasters states that the U.S. is at a serious disadvantage versus foreign competition. To alleviate the situation, he proposes that the U.S. Congress enact a set of specific pieces of legislation. He also suggests that a number of other unofficial measures be taken to create a balance in the foreign competition situation.

The final contribution by Richard E. Horner, Chairman and Chief Executive Officer of the E. F. Johnson Company, consists of a general overview of the state of the U.S. telecommunications industry with regard to international trade. It is very much a policy statement. Among other suggestions, Horner would like to see massive changes in U.S. tax policy. He would also like to see some coordination between tax and money-spending policies. Commercial banking is another area of concern for Horner, as well as a number of pieces of legislation that he thinks would help improve the U.S. situation.

The interesting point about all of the industry profiles and analyses in this book is that they all, to a greater or lesser extent, sound an alarm. There is growing concern within U.S. industry about the policies of foreign governments and, given these policies, the inability of U.S. industry to effectively compete in the world marketplace. There is also a tremendous amount of concern about the U.S. government's lack of initiative in international trade. Time and again, industry leaders can be heard complaining about the U.S. government's failure to support its key high technology industries. Instead, U.S. industry is left to compete unfairly with foreign industries that not only have policy support from their governments but financial support as well.

In addition to the identification of the problems that confront U.S. high technology industries is a snapshot of the state of high technology, at least as it pertains to semiconductors, computers, robots, and telecommunications. These descriptive profiles are of great value to anyone interested in learning more about U.S. high technology.

Our hope is that through the publication of these profiles and analyses we can contribute to a narrowing of the gap between those who make policy and those who must live with the results of ill-conceived policies. There is no question that the issues raised in this book are of tremendous importance to the future of the U.S. economy. Indeed, if many of the issues raised in this book are not attended to in the very near future, the economic future of the United States might very well remain threatened for decades to come. As many of the contributors to this book have repeatedly stated, the future is now.

The Editors of Petrocelli Books

Introduction

The High Technology Meetings

In the past two decades, economic development in the United States has increasingly depended upon high technology industries. The "high tech" sector has contributed significantly to economic expansion, employment opportunities, and national productivity. It has also become increasingly important to the competitiveness of other sectors. All indications are that its importance will continue and grow in the future.

The same trends are clear in other industrialized countries. Each in turn has recognized that its high technology capabilities may critically influence its long-term economic success. Out of this has grown an atmosphere of intensifying international competition in most high technology fields, and a keener awareness of the role that public policy plays in shaping and directing this competition.

In order to assess future problems in U.S. high technology competitiveness, the Secretary of Commerce held a series of meetings in January and February of 1983 attended by leading executives of four high technology growth industries and high-level Administration representatives. The four industries chosen were robotics, computers, semiconductors, and telecommunications. These four industries were selected because they are key sources of future growth and productivity for the U.S. economy. In addition, these industries all face significant problems associated with the "targeting" practices of foreign governments.

In his opening remarks, the Secretary stated that the meetings were designed "to explore the problems and challenges you face as an industry and to exchange views with you as to what the U.S. Government should—or should not—do in response to those challenges. We want to know what you see ahead and how you plan to respond as an industry. We also want to know if there are appropriate ways the government can facilitate your competitiveness."

The Department of Commerce found these meetings very useful. Important issues were discussed frankly. The Department regards the meetings as the beginning of a widened and improved process of communication between government and industry. The Department expects to arrange similar meetings with other industries.

This book contains four statements and discussions of the problems these four industries face, as they emerged from the meetings. We believe these issues deserve continued wide discussion by an informed and interested readership. Each Part is devoted to one of the four industries and contains two major sections: an overview of the industry, and a second section presenting the papers given by the industry speakers during the meetings.

Industry Profiles

The industry profiles present a brief assessment of the competitive situation faced by each of the four industries. Each profile is designed to:

1. assess the industry's international competitive position;
2. identify important competitive issues; and
3. present for discussion options to address these issues.

Presentations Made by Industry Speakers

The issues raised by the various industry speakers fell into two categories: first, issues common to more than one of the four industries and second, issues specific to the industry in question. A list of the general issues raised in the meetings as a group appears below, followed by an outline of the issues specific to the semiconductor industry. The industry presentations discuss the specific issues in greater detail.

General Issues Developed
from the High Technology Meetings

A total of eleven generic issues emerged from the four meetings. Each industry tended to rank the importance of the issues differently. The list below enumerates the issues raised during all the meetings in no order of priority:

■ Access to foreign markets.

■ Better U.S. Government response to foreign governments' targeting practices.

■ Use of fiscal and tax policy to provide incentives for high technology R&D and applications.

■ Dampening effect of present antitrust regulations.

■ Export controls and licensing.

■ Government support of research and development.

■ Formulation of a U.S. industrial strategy covering both the domestic and international markets.

■ Need for better technical/scientific education to ensure supply of qualified personnel.

■ More assertive U.S. role in multilateral and bilateral trade negotiations.

■ Better support for Eximbank.

■ Better export promotion.

INTRODUCTION

Each industry's speakers focused on the issues of principal concern to their own industry. Thus, not all of the eleven issues listed above were raised in each meeting.

U.S. Department of Commerce

1
The Semiconductor Industry

A Current Perspective on the Semiconductor Industry

PREPARED BY THE
U.S. DEPARTMENT OF COMMERCE

Introduction

Semiconductors are the building blocks for nearly all types of electronic equipment. As such they are a critical component in the technological progress of the world's industrialized countries. The integrated circuit, an advanced form of semiconductor, is a key to this progress. Virtually all major developed countries have recognized the importance of this industry to their future industrial development.

The domestic and international markets served by the semiconductor industry are extremely complex. Semiconductors feed a wide range of final markets: commercial-industrial, military, and consumer. In addition, each electronic component sector competes with other electronic component sectors. For example, integrated circuits have displaced other components, such as resistors, capacitors and transistors, in many types of equipment. The use of integrated circuits has also made possible the manufacture of many new products, such as personal computers and pocket calculators.

The semiconductor industry depends heavily on international trade. In 1982, exports accounted for 38.3 percent of the sector's product shipments. Imports as a percent of new supply (product shipments plus imports) increased between 1972 and 1982 from 12 to nearly 29 percent. In the same period, exports of semiconductor devices grew at a compound annual rate of 22.8 percent, while imports increased at 27.7 percent per year. The import growth reflects an increase in U.S. offshore operations and heightened competition from Japan, particularly in the random access memory (RAM) segment of the semiconductor industry.

The U.S. semiconductor industry, with domestic shipments in 1982 estimated at $9.5 billion and the continuous introduction of new devices, is the world's leader in both size and technology.

Foreign Competition and Markets

The Japanese semiconductor industry, our closest rival, has demonstrated that it can challenge the U.S. leadership position in at least some selected market areas. The recent Japanese penetration of the 16K and 64K random access memory (RAMs) markets in the United States has caused a great deal of concern among U.S. semiconductor manufacturers. These particular devices are high volume items and are important to the manufacturers' profit base. In addition, they represent critical steps in the development of semiconductor product and process technology. Failure of a company to compete effectively in any one of these markets can have serious consequences for its future development in such state-of-the-art devices. Japanese semiconductor manufacturers are well aware of the importance that large market shares play in fostering their future growth.

For example, the following statement appeared in the June 29, 1982, issue of the Japan Economic Journal:

SEMICONDUCTOR

"Semiconductors are very likely to determine the level of a country's computer, telecommunications, robotics, aerospace and other high technology industries in the future. The reason is that microchips now constitute the 'core' components of highly-sophisticated products. Some people call semiconductors 'the crude oil of the 1980s'."

Our semiconductor trade deficit with Japan has grown from $71 million in 1979 to an estimated $370 million in 1982. Much of this problem is due to the impeded access U.S. manufacturers face in the Japanese domestic semiconductor market, while Japanese producers operate freely in the U.S. market. In addition, there has been close government/industry cooperation in Japan which, our semiconductor companies believe, has been of significant benefit to the Japanese semiconductor industry. Japanese semiconductor producers have successfully penetrated the U.S. semiconductor market with high quality competitively priced devices made possible, to some degree, through government/industry cooperation and their operation from a protected domestic market.

U.S. manufacturers resorted to offshore operations in the early 1960's to improve their competitiveness in both their international and domestic markets. Industry leaders first established assembly facilities in foreign countries to take advantage of lower labor costs, or in the case of Europe, to hurdle extensive tariff and non-tariff barriers. When a large part of the offshore production began to return to the United States, other domestic semiconductor manufacturers were forced to follow suit. Today it is estimated that 80 percent of total U.S. semiconductor imports come from U.S. offshore manufacturing facilities and that a similar percentage of our exports are parts being sent outside the United States for assembly. Offshore manufacturers clearly play a significant role in semiconductor trade.

Fair and open trade in the European market is important to the U.S. industry because of the current size of this market and its long-term growth potential. At present U.S. manufacturers operating from European facilities hold a large share of that market and provide technological leadership. The European community is the major market for finished semiconductors, taking an estimated 43 percent of our exports of such devices in 1982. However, among the problems faced by U.S. industry in Europe are the high tariff barriers (17 percent duty) and non-tariff barriers such as the "Rules of Origin" which restrict U.S. devices from moving freely in the European countries without additional duties.

Current Situation

Aided by price declines, shipments, measured in 1972 dollars, rose 5.5 percent in 1982. In the 1972-82 period, shipments measured in constant dollars grew at a rate of 17.1 percent. Semiconductor prices have declined at an average rate of 1 percent for the past 10 years.

The economic downturn of 1982 also affected employment levels. For 1982, total U.S. employment in the industry rose only 0.7 percent to 168,300. Production worker employment fell 2.3 percent, to 82,200. Between 1972 and 1982, total employment grew at a rate of 5.6 percent, while the number of production workers increased 3.5 percent per year.

The ratio of production workers to all employees has fallen from 51.3 percent in 1974 to 39.9 percent during the first six months of 1982, as more expensive and sophisticated production and processing equipment was introduced. Between 1977 and 1980, capital expenditures almost quadrupled at the same time that product shipments doubled. Capital expenditures rose from 9.0 percent of product shipments in 1977 to 16.9 percent in 1980.

Depressed end-use markets in the United States and Europe were in large measure responsible for the semiconductor industry's lackluster performance during 1982. Growth in end-use demand in the computer, industrial-commercial and consumer markets remained well below pre-1981 levels.

Confronted with a two-year (1981-82) product shipments growth rate of less than one percent, the industry, especially during 1982, became hard-pressed to maintain capital expenditures and employment at pre-downturn levels. Industry strategy centered on keeping the Japanese share of the RAM market under 40 percent. Vivid memories remained of corporate decisions to slash capital expenditures and employment in 1974-75. Industry observers now see these 1974-75 moves as having opened the door to increased Japanese RAM imports. After the 1974-75 recession, U.S. demand increased so quickly that inadequate U.S. semiconductor capacity became a significant bottleneck. The Japanese had the excess capacity to fill the gap, and they supplied quality RAM parts.

Consequently, the 1981 downturn saw U.S. semiconductor manufacturers postpone capital expenditure cutbacks and personnel layoffs in order to be ready to compete effectively with the Japanese during the next cyclical upturn. However, in the first three quarters of 1982 the U.S. semiconductor industry resorted to both capital expenditure and employment cutbacks. In August 1982 alone, organizational realignments, layoffs or forced vacations affected some 1,500 employees. Although indicative of cost-cutting measures, the total number of affected employees was less than one percent of total industry employment.

During 1981, some industry analysts began to question whether the U.S. industry could compete effectively with the Japanese on the 64K dynamic RAM. In 1981, five Japanese and two American companies were producing in volume for the merchant 64K RAM market. By the end of 1982, three more U.S.-owned firms had entered the field. At the same time, the U.S. share of the world 64K RAM market increased from about 30 percent in 1981 to around 40 percent in 1982. In the U.S. market, American-owned companies increased their market share from 33 percent in 1981 to approximately 50 percent in 1982. Japanese volume producers included Fujitsu, Hitachi, Mitsubishi, NEC and Oki. The leading American challengers were Texas Instruments (TI) and Motorola. In fact, TI was the largest producer of 64K RAMs in the world during the second quarter of 1982.

The reasons the Japanese have concentrated so heavily on dynamic memories are straightforward. The 64K RAM is sold in extremely high unit and dollar volume, well suited for Japan's highly automated production facilities. Also, the future direction of dynamic RAMs is clear: a logical progression from the 1K to 1

7

megabit dynamic RAM. Finally, dynamic RAMs are one of the essential components for computers, a key product in Japan's apparent technological and industrial strategy for the 1980's.

By mid-1982, prices for 64K RAMs had stabilized at about $5.50-$7.00 per device, depending on device capabilities and volume. Production expansion during the second half of 1982 put downward pressure on 64K RAM prices. At the close of 1982, 64K RAMs sold for approximately $3.50-$5.00 per device. However, no major plant expansions occurred to meet the surge in demand; rather, U.S. manufacturers sought to utilize existing facilities.

The rapid decline of all RAM prices during 1981 and first quarter 1982 contrasted sharply with predictions made during early 1980 about the future price of RAMs. For example, 64K RAMs were projected to have an average selling price of $50 in 1981. In fact, the 1981-82 price was well below $10 per device.

With 16K RAM price under $1.00 and 64K RAM price under $7.00, U.S. producers became hard pressed to achieve acceptable profit levels in high-volume RAM shipments, especially with weakened end-use demand. Reduced profits during the first two quarters of 1982 resulted in subsequent declines in the stock market value of several publicly held semiconductor companies. By the third quarter of 1982, however, the industry picture had brightened to the point where semiconductor stocks were appreciating in value.

Between 1981 and 1982, takeover bids slowed considerably in the semiconductor industry. Two important motives for buying a semiconductor company—to obtain a captive facility to assure supply, or to make a profitable investment—just did not exist in a period when overcapacity and poor short-term profitability characterized many product lines within the industry.

Long-Term Prospects

Between 1982 and 1987, industry shipments of semiconductor devices, measured in 1972 dollars, are projected to grow at a rate of 16.0 percent to over $27 billion. Constant-dollar growth will continue to exceed current dollar growth as semiconductor prices decline.

Continued advances in both product and process technology will promote future growth of the U.S. semiconductor industry. Many of the newer devices, at the leading edge of technology only a few years ago, will find broad application by equipment producers. Devices such as microprocessors and high speed memories will continue to move into a wide range of commercial and consumer electronic products.

Further advances in manufacturing processes such as the electron beam (E-beam) and X-ray lithographic systems for semiconductor manufacturing are the key to future developments in very large scale integrated circuits (VLSIC) and very high speed integrated circuits (VHSIC). These production advances are expected to ease the manufacture of high density and high speed integrated circuits designed for use in the 1980's. As a result, further technological advances will occur in microprocessors, solid-state memories, advanced broadband communication systems, and guidance systems for missiles and satellites.

In the 1980's, foreign competition will intensify as nearly all countries will seek ways to develop and produce semiconductors to nurture industrial growth. In Japan, West Germany, France and the United Kingdom, government-sponsored semiconductor R&D will continue.

U.S. merchant semiconductor manufacturers will continue to confront two vexing problems throughout the 1980's. First, the United States will probably graduate proportionally fewer electronics engineers than Japan. Furthermore, U.S. engineers will continue to seek management positions much sooner than their Japanese counterparts. Consequently, it will remain extremely difficult to keep a top-flight U.S. design team together long enough to maintain essential continuity in a design project. Given the increasing complexity of future semiconductor devices, this continuity problem will become more troublesome.

Second, Japanese semiconductor manufacturers will, for the most part, continue to operate as divisions within large, highly integrated and diversified manufacturers. For major Japanese computer manufacturers, the return on their end-products will be more important then the return on any individual component.

Most U.S. semiconductor producers will remain in the RAM business, recognizing that a high level of memory R&D will mean benefits for other product areas where above-average returns on investment remain possible. These other product areas will involve a high level of innovation, shorter production runs, more software intensity, more systems orientation, and a need for close customer-vendor interaction. A first step may be the formation of operating units within major U.S. merchant semiconductor manufacturers concerned solely with producing custom integrated circuits to the exact, but often changing, specifications of prospective customers.

Semiconductor Industry Overview

BY ROBERT N. NOYCE
VICE CHAIRMAN
INTEL CORPORATION

As the President has said, there is a new hope appearing in our country—the hope of a productive economy without the inflation of the recent past, with both financial and physical security for all Americans. That hope is kindled for those of us in the semiconductor industry in part by the very fact that a meeting such as this is taking place, for we know the tremendous importance of technology for America's future. America needs to be positioned to take advantage of the enormous changes which are occurring in the world, and in our own society. Those changes are occurring largely as a result of technological progress, well exemplified by the advances made by the semiconductor industry. As you know, this relatively young industry, whose explosive growth has occurred only in the past ten

11

or fifteen years, is fundamental to the new information age. Half of the country's work force is now dealing with information rather than with goods. The advent of new information technology promises to improve the efficiency of that half of the work force even more than the mechanical age enhanced the output of manual labor in the last century. Even as building industrial America was the key to a rising standard of living, the building of America's knowledge industry will be the key to the continued enhancement of our national wealth and strength.

It is because of the seminal importance of that knowledge industry that it deserves your attention. Because of its primary role in our post-industrial society, this industry has been targeted by several governments as the engine to drive their economies to new levels of productivity. America's traditional role of technological leadership is being seriously challenged. If steps are not taken in the near-term to counter the structural advantages provided by these targeting practices, our country will miss a major opportunity to create new jobs, new wealth, and new security for the nation.

The semiconductor industry traces its origins back to the invention of the transistor shortly after the Second World War, but its major growth has come since the advent of the integrated circuit somewhat over twenty years ago. Because of that growth, we have seen the emergence of other problems. In the medium term, we will need significant new investment to provide for the plant and equipment for our growth, and, probably more significant, for the investment in research and development so necessary to remain competitive in this rapidly changing field. This problem has emerged in the last few years, coincidentally with the Japanese targeting of the semiconductor industry, since prior to that time our profits were large enough to finance our growth from retained earnings. With severe competition from Japan, earnings have been depressed to the point that they are no longer adequate to finance the growth necessary just to maintain market position. Furthermore, potential new investors are reluctant to provide the necessary capital so long as the Japanese targeting practices continue to be successful, for they question whether the American companies can survive that onslaught.

Another result of the rapid growth of our industry has been the shortage of trained personnel, whether technicians or PhD's, which is the basic determinant of success in any brain-intensive business. While the electronics industry tripled in the last decade, our nation's output of scientists and engineers has remained essentially constant. Japan is producing, on a per capita basis, nearly four times as many electrical engineers as is America. This lag in our support of technical education has had another undesirable side effect: the depletion of the common pool of research results and new technology from which we all draw. In a time of limited financial and human resources, we must find new ways of assuring that those resources available are most efficiently used. In short, we must usher in a new cooperation among industry, government, and academia, to provide the research and manpower base upon which to build this industry.

At the risk of validating the view of those who believe that American management has a short-sighted view, let me return to our short-term problem, i.e., that of government targeting. If this problem is not adequately addressed, our longer term problems will become, I fear, academic. In order to show the severity of this problem I would like to direct your attention to three facts:

First: the expenditures on research and development, which as I noted is one measure of the rate of technical advance, in the U.S. and Japan. For every U.S. dollar spent on R&D in 1977, Japan spent less than forty cents. Today that figure has risen to well over 70 cents and is still rising.

Second: the Japanese investment in the semiconductor industry has virtually overtaken that of U.S. merchant semiconductor firms. If present trends continue, they will overtake the investment of the U.S. merchant and captive firms combined. This investment has resulted, at least for now, in a capacity far in excess of that needed to supply domestic demand.

Third: partially as a result of that over-investment, Japan has launched an export drive, while effectively limiting imports. As a result, the trade balance with the U.S. has reversed, and now is heavily in Japan's favor—and becoming more so.

And it may well be that this critical industry in total is operating at a loss in both the U.S. and Japan today.

I would also like to address the subject of export controls and technology transfer. I would like to stress that we support reasonable controls to prevent unwarranted leakage of valuable know-how to the Soviet Bloc, but great care must be exercised to assure that the controls are exerted on a multilateral basis with our COCOM partners, and exports must be not unduly restricted. In this regard, we have recently responded to a proposal by Defense Under Secretary DeLauer to impose tight controls on IC process know-how and keystone equipment while liberalizing integrated circuit licensing procedures. If such a new system of control on technology is to be established, we believe existing commercial safeguards on technology must be recognized.

We have, therefore, proposed creating a *Comprehensive Operations License* system. This system will enable high technology firms which have demonstrated the adequacy of their internal controls to move equipment and know-how within their multinational network under a *Comprehensive Operations License*, precluding delays inherent in technology transfer.

I would like to offer one more perspective on our industry. Many experts have pointed out that the semiconductor industry is the foundation upon which the modern electronics industry is built, and the key to its competitiveness. The electronics industry is comparable to, or has surpassed, others which have received significant assistance from the federal government: agriculture, aircraft, and autos. (Perhaps we should rename the industry "automation" so that it starts with *A*.) There is a major difference between those industries and ours, however. The electronics industry has the potential of creating many new jobs for America, while at the same time contributing to our national goals of security, productivity, and quality of life for all America. It would be wise for us to nurture such an asset.

International Trade Policy

BY W. J. SANDERS, III
CHAIRMAN OF THE BOARD
PRESIDENT AND CHIEF EXECUTIVE OFFICER
ADVANCED MICRO DEVICES, INC.

United States industrialists have compared the effect of Japanese industrial strategy—in particular the demolitionary practice of industry "targeting"—to a rifle shot, exerting tremendous pressure on a very narrow area.

The Japanese, in contrast, have boldly stated—and I quote Hitachi's Components Operations, Executive Managing Director, Mr. Asano—"It's not a rifle, it's a cannon."

This disparity of perception dramatizes the fact that the critical industrial and government influences in this country have yet to realize the magnitude of the Japanese challenge and the intensity of the current trade conflict between U.S. semiconductor companies and our Japanese competitors.

Semiconductors are the basic building blocks of most high technology industries—industries which hold the key to this country's economic future. The health of the semiconductor industry is essential if the U.S. is to strengthen its domestic economy and export base during the remainder of this century. A healthy U.S. semiconductor industry is absolutely fundamental to improving this nation's productivity and international competitiveness.

Today we face what is likely the most serious challenge to our economic—and ultimately our defensive power—in recent history. In the mid-seventies, the Japanese Government determined that its firms should dominate the world market in certain critically important semiconductor product lines and launched a national effort to achieve this goal. U.S. semiconductor firms are now feeling the impact of that Japanese effort. U.S. firms have suffered substantial operating losses in the product lines targeted by Japan and are beginning to be hesitant toward making further investments in the areas which Japan has chosen to dominate. We are not talking about a special interest group here—we are talking about a vital American resource. And action must be taken now while U.S. semiconductor innovative and technical capacity is at its height—not only to save this critical resource from unfair competition, but to grow it.

Some Additional Background

In the sixties and early seventies, the U.S. semiconductor industry enjoyed a clear technological superiority over Japan. Japan permitted imports of U.S. semiconductor products which its own firms could not make and which Japanese and product firms needed to remain competitive. At the same time, and as an integral part of its industrial strategy, the Japanese Government protected its domestic semiconductor industry through import quotas and exclusion of foreign investment. This permitted Japanese semiconductor firms to develop technological capability, establish production facilities and build a dominant and secure domestic market position. Japan began restricting the import of overseas integrated circuits when similar ICs became domestically available. The U.S., then—in stark contrast to its leading position in the rest of the world—was limited to a

16

very weak position in the Japanese market. And that market is the second largest market in the world, second only to the United States.

In the mid-seventies, the powerful Japanese Ministry of International Trade and Industry (MITI), set a long range goal for Japan: world leadership in the high technology industries. Because of the importance of semiconductors to these industries, the promotion of Japan's semiconductor industry was a central element of this program. MITI "targeted" this industry for accelerated growth. The steps taken by MITI to promote Japan's semiconductor industry included government subsidized research and development; low interest loans; special tax advantages; the use of government facilities; protection from anti-monopoly laws; and free access to the results of government research and development. Between 1976 and 1982 Japan's major semiconductor firms have benefited by approximately 2 billion dollars worth of government support. Yet government funding is a small part of the real benefit to Japanese firms. More important, that funding provides a signal to the Japanese banks as to which industries are favored and are thus the best prospects for loans.

The business environment created by Japan's national targeting effort exhibits another critical element: the risks of large-scale capacity expansion and aggressive pricing are far less than the risks of failing to take such steps—the exact inverse of the situation here in the U.S.

Targeting policies have created another critical advantage for Japanese firms: joint research cooperatives subsidizing R&D and promoting shared technology among Japanese firms permit rapid entry into volume production. The well-documented 64K RAM experience of 1980 to 1982 is a prime example. U.S. firms had to develop their technology separately, resulting in a longer gestation from design to market.

With their early start, Japanese producers began cutting prices sharply. 64K RAM prices dropped 96 percent, from an average 25 to 30 dollars per unit to about 8 dollars during 1981. This dramatizes the sharp departure from the historic price curve caused by Japan's volume introduction of 64K RAM products. In addition, U.S.

firms' steepest 64K RAM operating losses coincided with the sharpest increase in Japanese shipments.

One Japanese semiconductor executive (Keiske Yawata, NEC Electronics U.S. Chief) remarked in an interview with *Business Week* (December 14, 1981), "I don't think anyone will be profitable at the projected (mid-1983 64K RAM) price of $5 for that period. The latecomers will be shaken out."

The resulting high losses experienced by U.S. firms can only serve to discourage future investment and participation in the RAM market. It's become a new ball game called, "You Bet Your Company", and the consequences of *disinvestment* are serious. RAMs traditionally have generated technological capability that translates into leadership in other semiconductor product areas. The RAM experience represents the "entering wedge" of an all-out assault on other semiconductor markets—repeating a pattern familiar to many other industries targeted by the Japanese.

It is also clear that the Japanese market is still effectively protected against foreign competition. U.S. sales as a percent of Japanese domestic semiconductor consumption, which were never large, are now declining and are actually lower than at points in the past when the market was officially protected. The market for 8080-type microprocessors collapsed in Japan while world demand continued. This collapse coincided with the introduction of volume production of 8080 product by the Japanese firm NEC.

The creation of an advantaged market environment through targeting incentives has fostered expanded capital investment by Japanese firms.

While the Japanese government covers the risk of aggressive investment and pricing for its firms, U.S. firms must fend for themselves in this high-stakes game. Japanese targeting policy has effectively increased the risks of U.S. firms.

The bottom line is that Japanese firms have increased semiconductor exports dramatically and, at the same time, have increased both U.S. and domestic market share.

This target industry policy has distorted the conditions of competition in world semiconductor markets and has enabled Japanese companies to challenge the technological superiority and economic viability of the U.S. industry. We are calling for the U.S. Government to respond to this targeting strategy and take immediate action to restore the conditions of competition that are essential to a healthy U.S. industry. Significantly, Japan's targeting policies and continuing protectionism have harmed U.S. firms in a way which entitles the U.S. to seek redress under international agreements which both Japan and the U.S. have signed—most notably, the General Agreement on Tariffs and Trade and the Subsidies Code negotiated in the Tokyo Round of Multilateral Trade Negotiations. We are recommending a six-point program of action:

First: The U.S. Government should announce as a U.S. policy that foreign industrial targeting practices will not be allowed to undermine U.S. technological and economic leadership in this critical industrial sector.

Second: The government should establish, in cooperation with industry, a monitoring system with respect to major commercial semiconductor product lines. This is now underway as a result of the High Technology Working Group efforts.

Third: The U.S. Government should insist that U.S. semiconductor firms receive commercial opportunities in Japan that are fully equivalent to those enjoyed by Japanese firms, *including* those firms favored by MITI.

Fourth: In order to establish free market competitive conditions internationally, the U.S. Government should promptly seek enforcement of Japan's obligations through consultations and other procedures available under GATT and the Subsidies Code, and should be prepared to exercise U.S. rights under such agreements if necessary.

Fifth: The government should use the implementation of these policies with respect to Japan as a model for dealing with target industry practices of other countries.

Sixth: The Congress should enact legislation that provides the authority and means necessary to ensure that our government can carry out the policies and measures outlined above. We recommend that the High Technology Trade Act of 1982 be reintroduced in Congress and that the Administration support the passage of this legislation.

Semiconductor

The U.S. companies affected recognize their responsibility to support these efforts and will do everything possible to meet these challenges in cooperation with our government.

Investment and Tax Policy Issues

BY GARY L. TOOKER
SENIOR VICE PRESIDENT
MOTOROLA, INC.

An aspect of targeting that has been inadequately recognized is its impact on risk. In the United States, there is a strong, direct relationship between risk and return. You expect a higher return when you invest in a higher risk opportunity. In Japan, the situation is quite different, as was illuminated two years ago when Chase Manhattan Bank did a study comparing the cost of capital and its return for the semiconductor industries in the U.S. and Japan. That study determined that, while U.S. semiconductor firms consistently returned more than their cost of capital, Japanese firms were able to operate over extended periods returning less than required to cover their cost of capital.* Further, their cost of capital was about one-half of ours due primarily to their heavier debt position.

* U.S. and Japanese Semiconductor Industries – A Financial Comparison, 6/9/80, pgs. 2.1-2.8 (Summary)

When a large U.S. semiconductor firm asked a Japanese banker why he would lend to the Japanese company that had a high debt to equity ratio and would not lend to him, with less than 25% debt, the reply was, "because I know I'll get paid by the Japanese firm."

Clearly, the Japanese government has taken the risk out of investing in target industries in Japan.

We don't propose that the U.S. Government take the risk out of any U.S. investments. We do ask that there be a full awareness of the competitive environment when considering tax measures that impact high technology industries.

Let me now address the financial problems of our industry and then get into the specific tax measures of concern to the SIA and how they will affect us.

Capital is the life blood of the U.S. semiconductor industry. To remain competitive, U.S. firms must invest at a much higher rate than most other U.S. industries. A typical merchant semiconductor firm spends 8-12% of sales on R&D, and 15-25% of sales on new plants and equipment. To maintain these high levels of investment, the industry needs to make a reasonable rate of return on current products, have confidence it can make adequate returns on future products, and be able to support substantial negative cash flow.

The emergence of the Japanese as major competitors due to their government's targeting policies has created a situation in which many producers cannot achieve or expect sufficient returns to stay in a growing number of product lines. In effect, we are disinvesting as a nation in certain key areas due to Japan's national policies. In the semiconductor Dynamic Random Access Memory, or DRAM, business we saw the number of U.S. firms in volume production fall from about 15 at the 4K level to 12 at the 16K level, when the effects of Japanese efforts began to have a significant impact. At the 64K level, only five U.S. producers remain, reflecting the early lead the Japanese firms got through their government's R&D programs. At the 256K level, there probably will be even fewer U.S. participants unless fundamental changes occur.

Some might argue that such attrition is normal as costs increase with each succeeding, and more complex, generation of products. There is some truth to this. However, that is only a partial explanation at best. If that were the primary reason, then how could the Japanese hold constant at six producers through three successive DRAM generations. Why did none of their firms drop out? The answer is they had *no* risk, *plenty* of borrowed money, and *no* pressure to make a reasonable profit.

On top of the clear problems in individual product areas, we can see a general pattern of investment levels in semiconductors that represent a broad threat to future U.S. industry competitiveness. The Japanese have created an environment in which capital expenditures increase rapidly, regardless of economic conditions. They go up rapidly during recession as well as expansion periods. In U.S. financial markets, it is difficult for firms to behave in this fashion.

The bottom line is that Japanese industrial policies have fundamentally altered the attractiveness of U.S. investment in certain targeted products. A prudent U.S. executive must have unusually strong incentive to remain in a targeted business and *must* have access to capital not received from current generations of such products. Such incentives and sources are unlikely to sustain an effort indefinitely.

SIA's study of the Japanese semiconductor industry found a variety of types of government financial assistance that have been and continue to be directed to the industry. These include grants, low interest loans, special depreciation and tax benefits, and export incentives. This assistance, coupled with other forms of support, has created a clear advantage for certain Japanese firms in international competition.

On our side of the ocean, we are faced with changes in depreciation under ACRS that benefit older industries, but actually were detrimental to high technology firms. With the 1982 modifications, the result is a reduction in tax benefit for most equipment at all realistic discount rates. In view of the rapidity of changes in some parts of our business, it would be highly desirable to alter depreciation allowances to accommodate short-lived equipment. Our recommendation would be to eliminate the five-year required life and to consider allowing expensing equipment in the year purchased. As you know, the R&D tax credit is of particular importance to our industry. This credit should be extended beyond its planned 1985 expiration and, hopefully, expanded to include all R&D employee costs. The law should also be modified to provide zero base credits for corporate contributions to university research and education.

Recent IRS interpretation of the R&D tax credit legislation excludes most software from the credit. The software development costs for some new VLSI products are greater than the IC development costs, and are an integral part of the product. In addition, major new software development projects are products themselves. Software R&D costs for engineering programs and new products should be specifically included under the R&D credit provisions.

Because most of our industry's firms are active internationally, including manufacturing offshore, U.S. tax policies on foreign activities are quite important to us. We understand GATT problems with DISC and we strongly agree with the Administration's position that any alternatives be, at a minimum, a tax neutral change. In addition, the Section 861 R&D reallocation rule should be reconsidered on a permanent basis.

In the current budget environment, we do not expect U.S. tax policy to provide a major boost for U.S. semiconductor firms—to offset the Japanese government's support for its firms.

The items just mentioned will not have a significant negative effect on federal revenues, but can have a very positive effect for companies in our industry. In particular, the R&D tax credit must be extended or our industry will face increased tax liabilities, making us less able to compete with foreign companies and less competitive for financial resources vis-à-vis other U.S. industries.

As the Administration considers tax issues with the current Congress, SIA believes it is imperative that the competitiveness of the U.S. semiconductor industry in the world market be kept in mind. Should the fiscal environment permit, we would ask that the administration and the Congress examine additional ways of reducing the tax burden on U.S. semiconductor firms. Until the cost of capital to the U.S. and Japanese producers is in better balance, U.S. firms will remain at a serious disadvantage.

As a final point, let me suggest one action that would be of financial benefit to U.S. producers—reduction of U.S. and Japan tariffs on semiconductors from 4.2% to zero. Most U.S. producers do packaging of their semiconductors offshore. As a result, the industry pays substantial duties on the value-added abroad. Elimination of duties would save the industry over $100 million annually that could be reinvested in the business. Efficiency would be improved for both government and industry by totally eliminating all paperwork on 806 and 807 filings and on duty drawback. Additional millions would be saved by not having to pay any duty into Japan. Since Japanese pricing in the U.S. seems totally unrelated to costs or exchange rate variations, we don't believe a 4.2% reduction in U.S. duties will impact their market share here. We want this change quickly and it should not be linked in any way to a reduction in E.C. tariffs.

In many ways the impact of the U.S. climate for investment is already visible. The U.S. semiconductor industry is very close to abdicating the 256K DRAM market. This could ultimately result in a total dominance of the world semiconductor market by the Japanese. Favorable tax treatment, as outlined above, will have a small, positive effect. It will take the sum total of many positions—including change in our trade policies, R&D cooperation and tax benefits—to turn the tide and retain our worldwide market leadership in semiconductors.

Research and Education

BY ERICH BLOCH
VICE PRESIDENT, TECHNICAL PERSONNEL
IBM CORPORATION

The Need for Research

During the last 30 years, the phenomenal growth and success of the semiconductor industry can be attributed to its imaginative pursuit of research and the availability of well-educated and experienced scientists and engineers. This research had led to the rapid generation of new, fast-paced technologies and new processes. In turn, these new processes have led to a continuous flow of new products.

The semiconductor industry, just like its companion, the computer industry, continues to be a rapidly changing R&D intensive endeavor.

Innovation and new research results are generated daily and the growth of the past will be overshadowed by the growth in the future, if the pursuit of research continues.

In order to cope with increasing competition in the world market, the semiconductor industry cannot relax; in fact, it must increase its efforts in research and development. At the same time, the research tasks are becoming more complex; more capital intensive; its lead-time is increasing and the shortages of sufficiently-trained man-power makes the staffing of needed projects difficult.

For all these reasons, some research efforts are beyond the afford-ability of many individual companies. The approach of the past and the present, where each company performs its independent re-search, causes much overlap and duplication of efforts, and is giv-ing way to joint cooperative efforts in the U.S.

While U.S. semiconductor companies have spent a greater amount of their sales dollar on this research compared to that of the total U.S. industry (10% for semiconductors and 3% for U.S. indus-try), this represents a smaller percentage than what is being allo-cated by the Japanese semiconductor industry (13-16% of sales).

It is important to realize, therefore, that competition from foreign companies in semiconductor research is just as intense as it is in semiconductor products.

Other Nations' R&D Efforts

Japan's program in semiconductor R&D has enabled leading Japa-nese semiconductor firms to draw even with the U.S. in many areas and even pull ahead of the U.S. in some specific endeavors. The re-sult has been an increase in the Japanese market share for semicon-ductors; especially in high-density memory products (64K dynamic RAM).

The Japanese effort has, and had, the participation and the guid-ance of its government. Government funding flowed into this effort; but equally important, participation of government scientists and re-searchers, especially from the MITI ETL Laboratory, contributed to the progress of their research. The sharing of the research results be-tween the participating companies accelerated the capabilities of these companies to compete in world markets, while sharing of the results did not diminish competition between these companies.

Another factor often overlooked in Japan's R&D effort is the role of NTT. Research in semiconductors and computers funded by NTT is sometimes performed internally, contracted to others or jointly executed.

NTT which does not have a manufacturing arm, but depends on major Japanese semiconductor, computer and telecommunication manufacturers for its product needs, has assisted in the design of key memory chips 16K, 64K and now the 256K chip. It has helped develop microprocessor products, making the technology available free to Japanese manufacturers.

The participation of national governments in semiconductor R&D is not unique to Japan. The European Community is in the process of setting up similar R&D efforts through its ESPRIT program and major individual countries are supplementing this supra-national project with efforts of their own; specifically England, France, Germany and Italy. A billion dollar effort is under discussion; the cost to be split between governments and industry.

The U.S. Semiconductor
Industry's Answer

It is encouraging that this Administration is committed to keeping America the technological leader of the world. A leading U.S. semiconductor and computer industry is a pre-requisite to this goal.

Faced with increasing competition from abroad and the realization that research is the foundation for its future well-being, the U.S. semiconductor industry a year ago established the Semiconductor Research Cooperative (SRC) to increase much needed research efforts, to share in its cost and its results, and thereby avoid redundancy of effort. Today, this cooperative is functioning. For 1982, there were close to five million dollars committed to this effort by participating companies.

The goals of the cooperative effort are threefold:

SEMICONDUCTOR

A. To increase semiconductor research through contracts with the nation's universities.

B. Through the contracts, to stimulate the participation of graduate students and faculty in semiconductor research; to attract more students to this field of study.

C. To upgrade and alleviate shortages of equipment and instruments in university laboratories.

It is our firm intent to increase in 1983 the industries' contribution and to double the 1982 level of funds. To accomplish this, we need to attract companies that are not participating today and we have excellent prospects in doing just that.

The industry cannot stop at that point, however, and must bring into these activities resources that are not readily available to the industry today. We had, and are continuing to have, discussions with some of the national laboratories, Los Alamos in particular, to tap its scientific and technical base for research projects that the SRC would fund, either alone or jointly with the national laboratories.

We also want to explore the participation of the government departments, such as DARPA or other areas of the defense community, to undertake jointly with the SRC and the academic community projects that are of mutual interest. This joint effort could benefit the VHSIC program. That program, a major undertaking, is providing leading edge semiconductor technology and products for defense. New research results can be used to maintain the necessary momentum for products needed for the national defense.

The Semiconductor Industry Association is, in addition, investigating the limited research partnership approach to leverage expansion of its research and development activities in product areas where individual companies cannot carry the task in both research and manufacturing by themselves. Another approach is a proposal to the Department of Defense to fund, together with industry, a "leap-frog Japan" effort in the area of multi-megabit memory chips.

While the Semiconductor Research Cooperative is focusing on research only, and is a non-profit organization, another activity between some computer and semiconductor manufacturers is the Microelectronics and Computer Corporation (MCC). (See Part 2 for more information on the MCC)

Scientific & Technical Education

As mentioned before, an adequate supply of qualified scientists and engineers is the life-blood of the semiconductor industry.

The reduction of government support to universities has forced industry and academia to rethink their relationship. This is resulting in new and positive actions, despite the fact that industry spends yearly about $240M on research in universities, this represents only about 4% of the total academic research budget. Industry can never fill the gap that continued withdrawal of Government funds would create.

The government needs to support research in universities and industry under a long-range strategic commitment, instead of the short-term "one year at a time" allocation. Research is by definition of long-range duration. Commitment to faculty and equipment procurement requires support over an extended time horizon.

As discussed before, continuing the R&D Tax Credit and zero-basing the credit for industry-funded university research is an initiative that Government should solidly support.

The changes to the Immigration Act proposed last year which would have closed the door for graduating aliens, engineers and scientists from entering U.S. industry would be disastrous to the future of high technology industry and would add to the current shortages of needed qualified people.

Summary

The future of the U.S. semiconductor industry depends to a great extent on its research foundation. The semiconductor industry has taken steps to:

A. increase expenditures for research;

B. share both the expense and the results of this research among the participating companies;

C. increase the utilization of U.S. universities; and

D. improve the quantity and quality of engineers and scientists that are produced by our educational system.

31

Industry's efforts could be greatly enhanced, if government would:

A. participate in the Semiconductor Research Cooperative;

B. stimulate research through liberalized and continuing tax credit;

C. increase its support to science and engineering in universities; and

D. through its policies and legislation have a long-range commitment to research for high technology.

Summary of Specific Issues Raised by the Semiconductor Industry

Summarizing the current issues of particular concern to the semiconductor industry:

■ The semiconductor industry believes that the Japanese semiconductor market is essentially closed to U.S. suppliers and wishes to see equal commercial opportunity in Japan for U.S. semiconductor sales.

■ Japanese targeting of the semiconductor market strongly affects the U.S. industry's competitive position and financial strength. The industry endorses the current agreements on monitoring semiconductors under the U.S.-Japan high technology agreements as a means of working out differences in semiconductor trade.

33

■ The effects of foreign targeting can in part be nullified through changes in tax credit and depreciation allowances for short-lived equipment.

■ Revisions to the antitrust regulations, particularly in the area of cooperative research and development, would allow the semiconductor industry to use research funds more economically and would reduce the uncertainties it currently faces in connection with potential antitrust suits.

■ The industry supports enhanced federal support for science and engineering in universities and would like to see a long-range government commitment to research and development.

■ The present shortage of adequately trained technicians and engineers will be reflected in a diminished capacity for industrial research and development. The semiconductor industry would like to see an increase in federal programs that will encourage engineering education. In addition, certain tax credits for industry grants to academic institutions should be made more liberal, thus ensuring funding and equipment from industrial sources.

2
The Computer Industry

A Study of the Competitive Position of the U.S. Computer Industry

BY ROBERT ECKELMANN
U.S. DEPARTMENT OF COMMERCE

Purpose and Summary

This profile is designed to:

1. assess the international competitive position of the U.S. computer industry;

2. pinpoint the major foreign and domestic challenges to American computer manufacturers; and

3. present for discussion possible options in terms of U.S. government policies affecting the sector's international standing.

The computer industry has become a cornerstone of the American economy. Over the last ten years, every major indicator in this sector—sales, production, and employment—has shown strong and consistent growth. On an international level, U.S. firms have occupied a position of overwhelming superiority, controlling some 75–80% of the world computer market during the 1970s, while watching the nation's annual trade surplus grow from $1.16 to $6.84 billion.

But recent trends indicate that U.S. dominance is being increasingly challenged, with the stiffest competition coming from Japan. While overall U.S. performance remains quite strong, Japanese computer development, accelerated by extensive industrial targeting programs, has progressed beyond simple control of their domestic market to a growing international presence. Though Japanese interests and activity span the full range of computer products, they have enjoyed particular early success in specialized segments of the industry. But other constraints, including more limited Japanese capabilities in software and services, have thus far prevented these inroads from being translated into a pattern of broad penetration. How long this will continue to be the case remains a subject of intense debate.

Last year, the severe global recession interrupted the historically impressive statistical performance of computer markets everywhere, but most analysts expect the lull to be brief and look to the rest of the decade as a pivotal period for American manufacturers. Their annual sales already exceed $65 billion, and forecasts indicate that 1990 could see that figure pass the $200 billion mark. This kind of continued success will require that U.S. firms:

1. meet unprecedented price competition across the entire range of computer products, both at home and in traditional export markets;

2. continue to pioneer new technologies while anticipating and absorbing advances in component industries;

3. maintain their leadership position in terms of the software and services that comprise a growing share of data processing revenues;

4. continue to parlay these advanced capabilities into a progressive systems approach to the broadening range of computer applications; and

5. expand aggressively into new foreign markets, situated primarily in the developing world.

The American computer industry is certainly capable of meeting these challenges, and by all accounts, should retain an impressive overall competitive position for the forseeable future.

But trade has steadily gained importance as a share of U.S. production, rising from 21.9% to 28.0% of output since 1972; at the same time, overseas subsidiaries remain a critical dimension of a healthy American computer industry. Therefore, as foreign programs more concertedly target computer development, and as a national computer capability evolves abroad into a more serious political, economic, and security concern, it will become imperative for private leaders and U.S. government officials to cooperate in:

1. assessing the magnitude and importance of whatever distortions might be introduced by such promotion;

2. seeking equitable access for American firms to the foreign markets involved; and

3. maintaining a U.S. policy stance that effectively incorporates the commercial interests, at home and abroad, of the American computer industry.

A number of options for possible USG action in response to the competitive challenges faced by the U.S. computer industry have been raised by various sources. These options are concerned with the following issues:

1. the U.S. response to foreign targeting practices that promote overseas competitors in computers;

2. USG policy on R&D, whether through public research, government support for academic and corporate activity, or tax treatment of R&D through the Economic Recovery Tax Act;

3. export controls on computer products;

4. the "skills shortage"—the problem of insufficient and inadequately trained manpower in computer fields.

Definitions

Due primarily to the rapid change that characterizes most high-technology fields, analysis of the computer sector presents a host of definitional problems. In terms of the area as a whole, the last few years have brought a gradual blurring of traditional distinctions between computers, telecommunications, and other "information industries", complicating even the simplest attempts to section off a discrete subject for study. Within the field, four powerful trends —dramatic improvements in computer capabilities, rapid evolution of their physical characteristics, steady expansion of their application and usage, and constant enhancement of their embodied price/performance ratios—have all necessitated constant modification of the standard labels used in computer product classification. Even then, the terminology always trails the marketplace. For purposes of simplicity and consistency, this analysis will adhere wherever possible to the following groupings:

Hardware	Mainframe	over $100,000
	Mini	$10,000 to $100,000
Software (and Services)	Micro	under $10,000
	Parts	full range
		full range

(The reader will be alerted to occasional situations where limited data and international discrepancies require the use of different categories and descriptions.[1])

The U.S. Market

Strong, steady growth has characterized the U.S. computer market over most of the last decade. Although not unaffected by the gyrations of the economy as a whole, the U.S. computer industry's fairly sustained performance remains the envy of most sectors. Since 1972, the overall rate of expansion has averaged 18.1% per

1. The other prevalent classification system employs a four-part breakdown - General Purpose, Minis, Small Business Computers (SBCs), and Desktops - in addition to Parts and Software and Services. The following diagram portrays the rough correspondence between these two structures and demonstrates the semantic problems inherent in this kind of analysis:

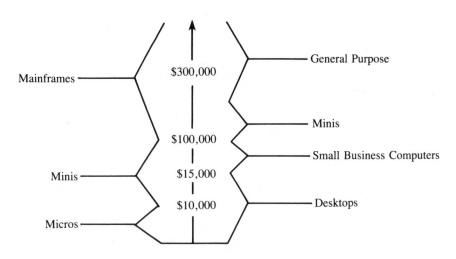

year (compounded, in current dollars), and despite the slight 1981–1982 decline to roughly 11.3%, most forecasters expect a 15–17% annual pace to resume at least through 1990. Also in 1982, employment growth slowed somewhat from previous norms, increasing only 5% over 1981 figures (versus the 12% achieved during the four previous years); but even this modest increase proved impressive against the general employment environment.[2]

Producers

Both Table 1 and Figure 1 summarize the solid overall record of the U.S. computer industry, as witnessed in years past and as expected in the future. But it indicates as well that aggregate figures disguise diverging trends for the major industry segments. Most striking is the sharp contrast between the large, mature mainframe market and the young, burgeoning microcomputer area—a clear signal that current patterns in technology, price, and usage relatively favor the low end of a still dynamic computer industry.

Recent strides in improving cost, capabilities, and convenience

2. Data refers to SIC 3573, "Computers and Parts", and originates in the *U.S. Industrial Outlook*, Department of Commerce.

Table 1
Worldwide Production of U.S. Computer Companies
(in billion $)

Total of which	1976	1981	1986
	23.4	56.0	108.3
Mainframes	6.2	17.2	24.8
Minicomputers	2.0	8.8	22.2
Microcomputers	—	1.2	3.5
Peripherals	10.0	13.9	18.7
Software/Services	5.2	14.9	39.1

Sources: Datamation; forecasts from survey of multiple sources.

have both opened up new mass markets for microcomputer products and brought unprecedented performance within full reach of a previously limited, mid-level business clientele. While the user side of the American market features an explosive diversification of demand, the producer side is led by the industry's largest manufacturer, International Business Machines Corporation. Yet its control over some 44% of U.S.-affiliated worldwide production (with an additional 14.6% of the domestic software and services market[3]) has in no way prevented other American firms from participating in the sector's vigorous expansion. Table 2 presents the overall sales figures, market share statistics, and annual growth rates for these leading companies.

The mainframers still comprise most of the "first division", but, as indicated in Table 2, firms specializing in mini and micro output are advancing rapidly. This pattern, an obvious corollary to the market trends noted in the previous section, becomes clearer in Table 3 and Diagram 1, which provide a breakdown of the competition within each of the main product areas. And since 1981 (the latest year for which hard data is available), the phenomenal growth and

3. Source: ICP Software Business Review.

Figure 1
Worldwide Production of
U.S. Computer Companies—1976, 1981, and 1986
(broken down by market segments)

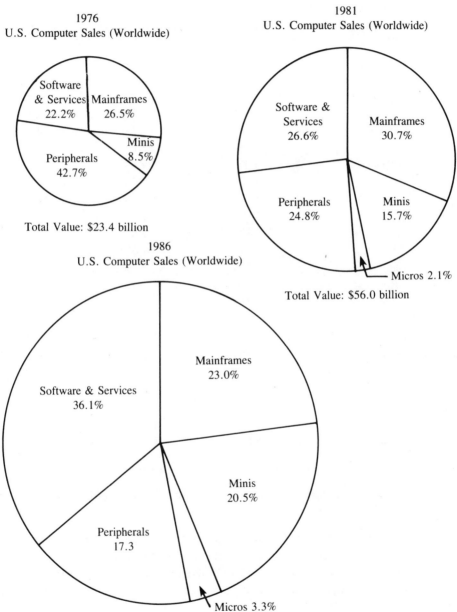

1976
U.S. Computer Sales (Worldwide)

Software & Services 22.2%
Mainframes 26.5%
Minis 8.5%
Peripherals 42.7%

Total Value: $23.4 billion

1981
U.S. Computer Sales (Worldwide)

Software & Services 26.6%
Mainframes 30.7%
Peripherals 24.8%
Minis 15.7%
Micros 2.1%

Total Value: $56.0 billion

1986
U.S. Computer Sales (Worldwide)

Mainframes 23.0%
Software & Services 36.1%
Minis 20.5%
Peripherals 17.3
Micros 3.3%

Total Value: $108.3 billion

Sources: Datamation, the Financial Times; forecasts developed from multiple sources.

COMPUTER

Table 2
The Leading U.S. Computer Companies–1981
(Worldwide Computer Revenues in $ Millions; Shares of U.S. Affiliated Production; Year-to-Year Increase, 1981 over 1980)

Firms	Revenues	Share	Growth
IBM	24,480	43.7%	16.6%
Digital	3,587	6.4%	30.7%
NCR	3,071	5.5%	4.1%
Control Data	2,893	5.2%	12.2%
Sperry	2,781	5.0%	8.9%
Burroughs	2,668	4.8%	24.6%
Honeywell	1,775	3.2%	8.5%
Hewlett-P	1,725	3.1%	18.4%
Xerox	967	1.7%	15.7%
Storage	922	1.6%	52.9%
TRW	815	1.5%	11.1%
Data General	764	1.4%	13.6%
GE	670	1.2%	57.8%
Texas Inst.	667	1.2%	6.7%
Compu. Sci.	625	1.1%	11.4%
ADP	613	1.1%	20.9%
ITT	484	0.9%	33.3%
Amdahl	417	0.7%	12.2%
Tandy	416	0.7%	109.0%
Apple	401	0.7%	142.7%
Wang	373	0.7%	47.9%
Industry Total (estimate)	58,500	100.0%	20.0%

Source:Datamation
Note: This table, which includes overseas revenues but not foreign competitors, does not reflect "market shares."

opportunity presented by the "low" end of the market has lured several of the top systems firms into across the board competition, matching their diversified strength against the prodigals of the personal and small business computer fields.

Table 3
Breakdown by Firm of the U.S. Industry's Worldwide Revenues
in Each Major Market Segment (1981, Millions of $)

Mainframes			Minis		
Firm	**Revenue**	**Share (%)**	**Firm**	**Revenue**	**Share (%)**
IBM	12,000	68.8	IBM	3,300	34.1
Burroughs	1,255	7.3	Digital	2,224	25.2
NCR	1,027	6.0	Burroughs	575	6.5
Sperry	918	5.3	Data General	573	6.5
Control Data	623	3.6	Hewlett-P.	435	4.9
Honeywell	511	3.2	Texas Inst.	320	3.6
Amdahl	335	2.0	Prime	309	3.5
Tandem	213	1.2	Honeywell	300	3.4
Natl.Adv.Sys.	175	1.0	Wang	272	3.1
Cray	102	0.6	Man. Assist.	244	2.5
TOTAL	17,200	(+9.3%)	TOTAL	8811	(+30.6)

Micros			Peripherals		
Firm	**Revenue**	**Share (%)**	**Firm**	**Revenue**	**Share (%)**
Apple	401	28.6	IBM	5,000	36.1
Tandy	293	20.9	Control Data	1,116	8.1
Hewlett-P	235	16.8	Sperry	1,112	8.0
Gould	140	10.0	NCR	1,015	7.3
Commodore	140	10.0	Storage Tech	786	5.7
Cado	68	4.9	Xerox	748	5.4
Cromenco	59	4.2	Hewlett-P	510	3.7
Total	1400	(+52.7%)	Digital	452	3.3
			ITT	400	2.9
			Textronix	309	2.2
			Total	13,850	(+10.8)

Software/Services

Firm	Revenue	Share (%)
IBM	4,480	28.0
Control Data	1,154	7.2
NCR	1,029	6.4
Digital	911	5.7
Burroughs	838	5.2
Honeywell	835	5.2
TRW	725	4.5
Sperry	695	4.3
Comp. Sci.	625	3.9
ADP	613	3.8
GE	570	3.6
Hewlett-P	545	3.4
Total	16,000(E)	(+26.0%)

Source: Datamation
Note: Software and Services are more narrowly defined in Datamation's survey than in the ICP Review (see footnote note 3). Hence the apparent incompatabilities in the numbers presented. Also, firms do not formally break out their revenues according to market segments such as these; therefore, in a strict sense, the above data should be regarded as estimates.

The World Market

Computer Hardware

On a global level, the computer market has had a healthy performance similar to that witnessed in the United States, with annual compound growth in the domestic output of all major producer nations averaging 20.3% over the last four years (1978–1981). The result: total production of computing equipment (SIC 3573) passed the $50 billion mark in 1981, and most forecasters remain bullish about the decade ahead, predicting that figure will exceed $185 billion by 1990. Table 4 denotes the present position of the six leading producer countries, according to output value, individual growth rates, and market shares.

More detailed analysis reveals that the product and sales trends which have come to shape the U.S. market also prevail on the inter-

46

Table 4
World Computer (SIC 3573) Production by Country

	1981 Output Value (billion $)	Growth Rate (1978–1981)	World Market Share
United States	29.53	23.2%	57.7%
Japan	6.70	17.5%	13.1%
France	4.88	18.1%	9.5%
West Germany	3.50	13.3%	6.8%
Great Britain	2.33	12.2%	4.6%
Italy	1.19	30.4%	2.3%
Others	3.07	—	6.0%
Total	51.20	20.6%	100.0%

Source: U.S. Industrial Outlook, 1983.

national scene. Again a pronounced trend is the upsurge of the small computer sector. Table 5 lists the seventeen most prominent companies worldwide. Within this group, mini and micro producers managed to raise their share of total data processing revenues by over 40% since 1978.

Software and Services

Through the 1960s and 1970s, software and services came to comprise an ever larger portion of data processing costs. For a typical mainframe system, this fraction now ranges from ⅓ to ½, and while the figure has recently shown signs of steadying, the fact remains that this dimension of computer products can decisively influence their technological and commercial success. As a result, international competitiveness in computers has become strongly correlated with a nation's software, as well as hardware, capabilities.

Over the past few years, the U.S. position in this area has seemed particularly strong; but as software has assumed a more central role in determining the performance and marketability of computers, the struggle for software leadership has intensified. Though technological change here does not lend itself to the simple performance summaries that chronicle hardware development, the unprecedented at-

Table 5
The World's Top Computer Firms
(Ranked by Computer/DP Revenues in billion $)

1979		1981	
IBM	18.24	IBM	24.48
Burroughs	2.43	DEC	3.59
NCR	2.40	NCR	3.07
CDC	2.27	CDC	2.89
Sperry	2.27	Sperry	2.78
DEC	2.03	Burroughs	2.67
Fujitsu (Japan)	1.49	Fujitsu (Japan)	2.03
Honeywell	1.45	Honeywell	1.77
CII-HB (France)	1.22	Hewlett Packard	1.73
ICL (U.K)	1.09	NEC (Japan)	1.51
Hewlett Packard	1.03	ICL (U.K.)	1.44
Hitachi (Japan)	0.98	CII-HB (France)	1.34
Olivetti (Italy)	0.98	Hitachi (Japan)	1.31
NEC (Japan)	0.91	Olivetti (Italy)	1.09
Siemens (FRG)	0.91	Xerox	0.97
Philips (Neth)	0.75	Nixdorf (FRG)	0.89
Nixdorf (FRG)	0.65	Siemens (FRG)	0.84

Sources: Corporate Financial Reports; Datamation; Bureau of Industrial Economics,
Department of Commerce; and others.

tention that the segment now receives is reflected both in the rapid domestic rise of software and services firms (Tables 2 & 3) and in the growing international competition for software markets. A summary of S&S activity in Europe and the U.S. (Table 6) reflects the continued expansion of recent years in both places. And the Japanese, supported by government programs that acknowledge software's indispensability, are attempting to improve their position in this field as well.

Software production has proven an essential ingredient of any viable computer industry. Its importance for the future is difficult to overstate. Certain keys have surfaced as essential to software success—high-quality, error-free products, standardization in pro-

Table 6
Europe and the U.S.: Software and Services

	1981 Market Size (mil $)	Current Annual Growth Rate (%)
France	1800	19.5
United Kingdom	1450	13.4
West Germany	1180	10.6
Italy	849	17.8
Netherlands	536	14.8
Sweden	444	11.3
Denmark	328	12.0
Belgium	325	14.0
Switzerland	300	11.8
Norway	268	21.2
Spain	241	22.2
Finland	236	15.6
Portugal	23	35.9
Western Europe	8170	15.3
United States	12500	14.8
U.S. firms' worldwide S&S revenues: $22.6 billion (1981)		

Sources: ADAPSO, Financial Times (Computer Services Outlook).

grams and languages, and custom capabilities (i.e., tailoring to specific applications). And in a field filled not only with independent houses but with systems firms that purvey packaged products, the continued adoption and application of creative new concepts will be required of those who are to emerge as leaders.

New Markets

The world market has gradually accepted the full range of computer products and activities; but from a geographic point of view, the most dramatic change has been its expansion to include a wider range of customers and users. Industrializing countries (led by Brazil, Mexico, and the East Asian NICs) have already demonstrated enormous potential as a source of future demand, with selected growth rates often exceeding 25% per year.

World Computer Trade

World trade in computer products has grown at a faster rate than computer production itself—overseas shipments, for each of the major supplier nations, have comprised a steadily rising share of both total output and consumption. The chart below provides an overview of this trend as it has evolved over the last few years.

The basic reason for this development is the continuing "internationalization" of the computer market as a whole. In terms of demand, the three main forces at work are its rapid rise in key areas of the developing world, its steady diversification in traditional but unsaturated industrialized markets, and the increasing overseas activities of American subsidiaries. In terms of supply, the primary consideration is the improving competitiveness of non-American

Table 7
Trade as a Share of Computer Production/Consumption
(for the major supplier nations)

| | 1978 | | 1981 | |
	Export %age	Import %age	Export %age	Import %age
United States	26.5	6.1	28.8	7.3
Japan	11.5	12.1	18.0	14.7
France	32.4	32.6	34.7	39.5
West Germany	51.2	55.4	66.5	69.5
United Kingdom	77.3	81.8	83.5	85.3
Italy	57.5	68.1	74.7	81.0
TOTALS	30.8	23.4	34.4	25.6

Source: U.S. Industrial Outlook, 1983.
Note: Exports are as a percentage of domestic production; imports are as a percentage of apparent consumption. (SIC 3573 only)

sources.* For the U.S. this has meant a rapid rise in imports which, in 1982, reduced the U.S. trade surplus by nearly $105 million. The overall U.S. trade position in computers (1978–1981) is summarized in the following Figure 2 and Table 8, which include a breakdown of the major sources and destinations of these international product flows (SIC 3573 only).

Obviously, the U.S. trade surplus remains quite strong. But its unexpected deterioration in 1982 may imply more than just currency movements and disproportionate softening of overseas demand in the face of global recession. It has also served to underscore the serious concern over foreign targeting of computer development and government intervention in high-technology trade. The country by country analysis which follows will cover each of the primary producer nations (and thereby each of our major trading partners), including assessments of each domestic industry's competitiveness. This analysis constructs the complex network of public and private international challenges that face American manufacturers.

Japan

Both the Japanese market and the Japanese industry have performed impressively over the last several years. Recent growth in production has averaged 17.5% annually (compound rates in U.S. dollars), fueled in part by a continuing surge in exports; consumption has expanded at the slightly more modest pace of 15.8% per year. Overall its posted 1981 output of $6.7 billion in computing equipment now places Japan securely in the runner-up spot in this category, and the established group of Japanese manufacturers (Fujitsu, Hitachi, NEC, Toshiba, Mitsubishi, and Oki) seems well-positioned for the 1980s. Their current standing and the details of the Japanese market are provided below (See Tables 9 & 10).

In 1979, a domestic Japanese manufacturer managed to displace IBM Japan from its top spot in their domestic market.[4] Now, 1981

---

* Of course, one could also consider foreign subsidiary activities as a supply-side force in both their production and re-export roles.

4. Fujitsu's total revenues for JFY1979 exceeded IBM Japan's figure for calendar year 1979.

Figure 2
U.S. Computer Trade (SIC 3573 only)
Imports by Source; Exports by Destination

Shares of U.S. Imports (1982)

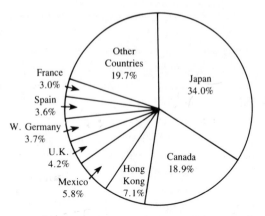

Total value of 1982 Imports: $2.14 billion
(SIC 3573 only)

U.S. Exports Shares, by Destination (1982)

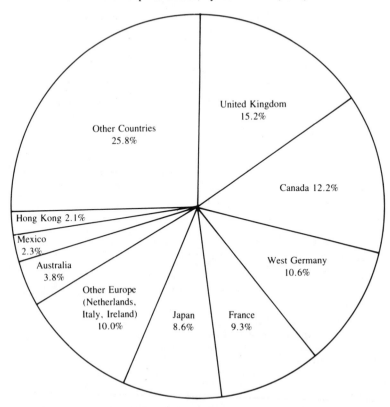

Total Value of 1982 Exports: $9.04 billion
(SIC 3573 only)

Source: Official U.S. Trade Statistics and Bureau of Industrial Economics estimates.

Table 8
U.S. Computer Trade (SIC 3573):
Origins and Destinations, Flow Value, and Annual Growth (1981–2, % of value)

1982 Imports ($mil)

	$mil	Growth
Total:	2140	(+ 29.9%)
of which:		
Japan	729	(+ 88.2%)
Canada	404	(+ 0.0%)
Hong Kong	151	(− 21.6%)
Mexico	123	(+ 27.0%)
U.K.	90	(+ 12.2%)
W. Germany	79	(+ 12.9%)
Spain	78	(+ 25.3%)
France	64	(− 2.6%)

United States

	Imports	Exports
1978	755	4194
1981	1647	8652
1982	2140	9040
Growth (78–82)	+ 29.8%	+ 21.2%

1982 Exports ($mil)

	$mil	Growth
Total:	9040	(+ 4.5%)
of which:		
U.K.	1374	(+ 15.3%)
Canada	1103	(− 11.2%)
W. Germany	958	(− 6.2%)
France	841	(+ 7.0%)
Japan	777	(+ 8.3%)
Netherlands	380	(+ 14.0%)
Australia	344	(+ 0.0%)
Italy	298	(− 4.1%)

Table 9
Computer Sales of 8 Major Computer Manufacturers in Japan
(Millions of Current Dollars)

	FY 1976	FY 1979	FY 1981	Average Annual Growth Rate 1976–1981 (%)	FY 1982 (Est.)	Chg. 1981–1982 (%)
Fujitsu	1,086.5	1,481.9	2,033.3	13.4	2,426.0	19.3
NEC	516.9	910.1	1,507.7	23.9	1,766.2	17.2
Hitachi	643.9	979.5	1,305.9	15.2	1,496.4	14.6
Oki	219.0	284.8	494.7	17.7	594.0	20.1
Toshiba	268.4	228.5	430.8	9.9	521.5	21.1
Mitsubishi	145.1	240.3	331.0	17.9	399.0	20.5
Major 6 Japanese Firms (1)	2,879.8	4,125.1	6,103.4	16.2	7,203.1	18.0
IBM Japan (2)	1,248.8	1,470.1	1,944.9	9.3	N.A.	N.A.
Nippon Univac (1)	320.1	333.7	412.2	5.2	453.5	10.2
Total 2 U.S. affiliates	1,568.9	1,803.8	2,357.1	8.5	N.A.	N.A.

Sources: Japan Economic Journal, June 9, 1981 and June 8, 1982, and the Bureau of Industrial Economics. 1981 exchange rate of 220.53 Y per $1 used for all years.

(1) Fiscal year ending March 31st.
(2) Fiscal year ending December 31st.

Table 10
Japan's Installed Base (1982)

Computer Size	# of Systems Installed	Value of Installed Base	% of Total Value of Instl'd Base	Growth Rate (of Value)
Very Large (over $2 mil)	1,914	$8.73 billion	44.4%	11%
Large ($1–2 mil)	1,586	$2.43 billion	12.4%	10%
Mid ($0.16–1 mil)	11,130	$4.56 billion	23.2%	16%
Small ($41–166 K)	32,565	$2.48 billion	12.6%	22%
V small (under $41K)	59,149	$1.45 billion	7.4%	20%

Computer Usage in Japan
(installed base, share of value)

Universities/Academics	16%	Elec. Machinery	11%
Finance	15%	Transport Machinery	4%
Distribution	15%	Chem/PetroChem	4%
Government	13%	Insurance	3%
		Others	19%

Source: MITI

has produced another milestone—for the first time, Japan posted a surplus in computer trade. Table 11 provides a breakdown of Japanese computer exports by firm.

It appears that the next objectives of the Japanese industry will be two-fold: to compete vigorously in the European market (where foreign penetration is already high), and to develop a firm footing in the U.S. market. Already, their presence is being felt in certain peripherals areas (such as small printers, disk drives, and auxiliary storage), is expected anytime in the micromarket, and should develop soon in the prestigious supercomputer field (where both Fujitsu and Hitachi have announced certain machine capabilities which

Table 11
Japanese Computer Exports by Company
(In Millions of Current Dollars)[5]

Company	1981	Change(%) 1980–81	1982(Est)	Change(%) 1981–82
IBM Japan	490	+ 57	N.A.	—
Japanese Firms[6]				
Fujitsu	263	+ 57	363	+38
NEC	172	+124	231	+34
Hitachi	132	+ 85	159	+21
Oki	59	+ 97	91	+61
Toshiba	36	+ 74	54	+44
Mitsubishi	32	+ 40	51	+29
Six-firm Total	694	+ 78	939	+35

Source: Compiled by the Bureau of Industrial Economics

they claim are beyond those currently available from their main American competitors, Cray and CDC).

Most forecasters anticipate that a Japanese growth rate near 20% is sustainable through 1985, resulting in a total shipment value by then of over $13 billion, a 10% share of the expected world market. That kind of success will doubtless require significant penetration of the U.S. market over and above full exploitation of their European potential. The precise timing of such developments is always subject to debate, but 1982 brought an 88% rise in computer exports to the U.S. and the first full results of several OEM agreements with European firms (Siemens, ICL, BASF, Olivetti).

The Role of the Japanese Government in Computer Development

Perhaps the greatest future concern surrounds the eventual impact of the Japanese Government's computer promotion efforts. Histori-

5. 1981 exchange rate: 220.53Y = $1
6. Fiscal years ending March 31, 1982 and 1983, respectively

cally, public policies have provided a broad range of support to ensure the continuing growth and competitiveness of the domestic computer industry. Most recently, these included (1981):

A. Tax Measures

1. Accelerated depreciation of computer purchases—an additional 13% first year write-off is permitted on all "machine types for the provision of industrialization";

2. 20% of total computer purchases can be deducted for purposes of local asset tax valuation;

3. Accelerated depreciation for computer producers—one-third of the initial book value of facilities used in the production of MITI-approved "newly developed technologies" is permitted as an additional first year write-off;

4. Tax deductions for computer producers:

a. 25% of all year-to-year increases in R&D expenditures (up to 10% of taxable income);

b. 50% of "software income realized" can be set up as a tax-free reserve to cover future software development costs;

c. 20% of all year-to-year increases in training costs for software engineers;

d. up to 2.5% of sales if placed in a reserve fund to protect against losses "caused by return of computers" via JECC, the joint leasing company.

B. Direct funding for major research programs:

1. Y100 billion over 8 years for 5th generation computer development;

2. Y25–30 billion over 9 years for super computer development;

3. Y23.5 billion over 5 years for "next-generation" computer research through the Fundamental Computer Technology Research Association;

4. Y20.0 billion over 7 years for research into "optical telemetering" technologies.

C. Government loans for leasing organizations (Y46 Billion in FY1981 to the JECC) and joint software programs (Y5 billion via the IPA Trust Fund and Long-Term Credit Banks).

Industrial policy support in Japan is well-managed, highly di-

57

rected, and efficiently funded. It has contributed to the rise of an internationally competitive computer sector and is now turning its attention to more innovative efforts. Though it is impossible to determine what impact current projects will have, recent history indicates that the Japanese challenge should not be taken lightly, and that distortionary public assistance could affect the U.S. share of world computer markets.

Western Europe

Western Europe represents the largest market[7] for computer products outside of the United States, and thus is critically important to the American computer industry. The U.S. sector's fundamental dependence upon tolerant, if not hospitable, treatment in Europe is reflected in its three-pronged involvement in the European marketplace:

1. through direct exports (the 1980 U.S. computer trade surplus with Western Europe was $4.1 billion);

2. through production by American subsidiaries for domestic consumption (in each of the major supplier nations, these firms accounted for over 50% of total output, for an overall estimate of $10–12 billion annually);[8] and

3. through trade within Europe among those same multinational enterprises (no precise figures are available, but over ½ of all European production is traded).

The following overviews of the destinations for exports of the principal competitor nations (Table 12) and the top European computer firms (Table 13) emphasize Europe's central role in the world computer market.

The basic characteristics of the European producer nations are:

1. the important role therein of U.S. subsidiaries;

2. the high traded fraction of total consumption and production;

3. the concentration of most national production into one or two major firms;

7. In 1981, demand in the U.S. market (SIC 3573 and separate software) was $25.7 billion. A comparable figure for Europe: $15.8 billion.
8. Auerbach; Foreign Trade News, 8/31/82.

Table 12
1980 Computer Exports of Principal Supplier Nations by Region
(In Millions of Current Dollars)

Destination by Region	Principal Supplier Nations					
	U.S.	West Germany	U.K.	France	Italy	Japan
North America(US/Can)	755	104	168	144	14	271
Latin America	615	29	7	29	2	61
Europe[9]	4,527	2,036	1,625	1,262	518	184
Africa	128	47	72	85	—	4
Asia/S.E. Asia[10]	1,382	32	36	43	4	170
Middle East	117	41	39	56	4	8
Communist Nations[11]	76	51	23	59	10	42
Other	6	7	211	14	318	1
Total	7,606	2,347	2,181	1,692	870	741

Sources: Official trade publications of each nation, compiled by Bureau of Industrial Economics, Science and Electronics Div.

4. the gradual encroachment of Japanese firms (especially Fujitsu and Hitachi) through joint ventures and OEM agreements; and,

5. the persistent attempts, through targeting programs, to promote the rise of a fully competitive national computer industry.

Even on an international level, certain European countries have sought to cooperate with one another in their attempts to challenge the U.S. and Japan in the world computer market. A major recent effort has been the European Economic Community's "Esprit" project—the European Strategic Programme of Research in Information Technology. A dozen firms from 5 countries will collaborate in both the current two-year pilot phase (begun in July 1982) and the eventual full-scale program, to start in 1984. The topics under consideration for study range from chips to 5th generation computers,

9. Includes Scandanavia and Greece
10. Includes Australia and New Zealand
11. Includes Peoples Republic of China

59

Table 13
Europe's Largest Computer Manufacturers (1981)

Company	Parent Company HQ	European DP Revenues (mil$)
IBM	United States	8,846
CII-Honeywell Bull	France	1,311
Siemens	West Germany	1,296
Digital Equipment	United States	1,162
ICL	United Kingdom	1,067
Olivetti	Italy	1,006
Sperry Univac	United States	850
Control Data	United States	765
Phillips	Netherlands	750
Burroughs	United States	742
NCR	United States	728
Nixdorf	West Germany	678
Hewlett-Packard	United States	604
CIT-Alcatel	France	556
Honeywell Inf Systems	United States	497

Source: Foreign Trade News, 8/31/82.
Note: Data not diretly comparable to that elsewhere in this report because it includes some "non-computer" revenues from word processing, data communications services, etc.

and the Community's technologists have estimated that its funding of the program in the late-1980s could surpass $2 billion.[12]

In the future, should European governments (either collectively or individually) voice increasing concern over the American presence in their computer sectors, and should the limited success of current targeting efforts then give rise to unacceptable forms of market interference, an important group of U.S. economic interests could be at stake. The following overview of the French, German, and British situations provides basic information on the current and expected market conditions in each country.

12. Financial Times, 8/3/82.

France

In the 1970s, France established itself as a European leader among computer-producing nations. But while 1981 output figures reached $4.87 billion, an unsettled economic and political atmosphere has limited French growth prospects in the eyes of most analysts, with typical forecasts hovering in the 7%–9% range. The forced rearrangement of CII-Honeywell Bull, the nationalization of Thomson, and apprehension over additional policy changes with the shift in government made for a more sputtering and hesitant year than originally anticipated. Even a generous program of industrial policy support has failed to restore fully the optimism that once prevailed. Several types of assistance are now in place:

1. preferential public procurement from national sources, as epitomized in French manufacturers' 63% share of the civil service's installed base (vs. a private market share near 45%);[13]

2. recent establishment of a "Super-Ministry" for Industry and Research (in imitation of Japan's MITI), with ambitious investment plans for France's electronic industries and with goals involving extensive technological cooperation with other European countries;

3. the "Farnoux plan", which intends to bring together private, nationalized and public sectors (including scientists, businessmen, users, and union members) in "national projects". By 1990, the proposed program hopes to double the French electronics market and increase total research outlays by 50%. In the meanwhile, it recommends a 3-year manpower effort, which would include the establishment of advanced electronics schools within existing technology-oriented institutes, the training of 2,000 top engineers (with an additional 10,000 technicians), and the gradual joining of the research efforts of Thomson, the PTT, and the French Radio and Television Office. Estimates of the cost of carrying out this segment of the program (including both public and private contributions) have run as high as FF10 billion. The scheme was developed in preparation for the French Government's new 5-year Microelectronics Plan.

13. The Market Monitor, 1982.

4. the minimization of competition between French firms, focusing on cooperation between CII-HB and Thomson (through, for example, the research and development stages of new minicomputer projects).

But although most observers remain skeptical of the benefits to accrue from this comprehensive package, it may presage more serious market intervention that could disrupt U.S. subsidiary activities and/or Franco-American computer trade. Consistent with the trade pattern outlined previously as typical of European nations, much of French output and consumption flows through the foreign sector. A fairly small 1981 deficit of $386 million (hardware only) occurred despite penetration rates of over 42.6% on imports of $2.1 billion; at the same time, exports of $1.69 billion accounted for nearly 35% of total production. Furthermore, the majority of domestic output falls to either IBM's French subsidiary (with a 50–55% estimated market share) or CII-Honeywell Bull (now less than 20% owned by Honeywell Inc., and with a 25–30% market share).

West Germany

The German computer market is the second largest in Europe, and the domestic industry includes two of the continents most competitive firms—Siemens and Nixdorf. Total national computer production (SIC 3573) in 1981 reached $3.5 billion, with these native non-subsidiaries performing as follows:

	Computer Revenues	% Growth over 1980	Estimated Market Shares
Siemens	841	20%	21.0%
Nixdorf	885	24%	22.0%

Note: Revenue figures for Siemens and Nixdorf, presented in millions of $, were adjusted to cover only computer revenues as defined for purposes of this report (i.e., excludes word processing, etc.).
Exchange rate used: 2.26 DM/$

As through much of Europe, the U.S. plays an important role in the German market—through direct exports, shipments from non-German subsidiaries, and the internal production of American firms. But Japanese competition has recently increased through OEM agreements such as those between Fujitsu and Siemens and between Hitachi and BASF. This development should soon manifest itself in Germany's trade figures, where like France, it maintains a persistent deficit (with nearly ½ of its imports coming from U.S. subsidiaries in the rest of Europe). Most forecasters also anticipate that currently modest growth figures (8–11%) will continue through most of the decade ahead.

The government's concern over the long-term future of the industry prompted the establishment, in 1977, of a significant network of support programs, which covered twelve financial plans in the three main areas deemed crucial to future computer development:

1. industrial R&D;
2. data processing applications; and,
3. manpower training.

Total disbursements for the effort (1977–80) reached DM 1.25 billion ($625 million at 2DM/$). Since its conclusion, most German government support for computer-related development has taken the form of R&D assistance to firms and universities through the Federal Research Ministry's Technology Center in Berlin. Some 266 million DM ($133 million at 2DM/$) will be provided in 1982 for its two major programs, in microelectronics and optic communications engineering.[14]

All in all, the German industry is perhaps the most competitive in Europe, but its relatively open market has enabled American firms to obtain a market share comparable to that common in more protectionist countries. Accordingly, though demand growth in their domestic market should offer no special promise, German computer manufacturers may provide an occasional challenge in the years ahead on an international level.

United Kingdom
The U.K. is home to ICL, the largest European computer manufacturer, with 1981 data processing revenues of $1.44 billion. However, expectations as to Britain's long-term growth possibilities and

14. VDI Nachrichten, Dusseldorf, 12/11/81.

competitive prospects are only mildly optimistic. Its total hardware market passed \$2 billion in 1980 and should expand at a 10–12% annual pace during the next several years. Growth in terms of installed base, however, does remain brisk:

Unit Value	% increase in 1981
above £30,000	+25.7%
£15,000–£30,000	+30.0%
below £15,000	+38.9%

Source: BIS—Pedder Census, 1981.

Apart from ICL, holding a 25–30% share of the home market, the main supply factors will be the American presence in the U.K. (with a 50–60% share, half of which can be traced to IBM) and the arrival of Japanese firms through OEM agreements. These arrangements, because they can include technology transfers (as in the ICL-Fujitsu case), could have long-term implications for the nation's production structure. The traditional British strengths have been software and services, and their tie-ups with Japan could provide complementary coverage of all aspects of computer systems.

In the meanwhile, government promotion of the computer industry has taken a variety of forms—grants, subsidies, loans, and publicly supported research:[15]

1. The Department of Industry has undertaken extensive funding (£80 million over 4 years) of "information technology developments". In a program to be administered through three existing frameworks (the Product and Process Development Scheme, the Electronics and Avionics Requirements Board, and the Software Products Scheme of the National Computing Centre), the government will directly support public research activity while furnishing grants to both firms and users to encourage more extensive computer application and usage.

2. Also proposed to the Industry Department is a joint research program aimed at development of advanced technologies for 5th

15. The Financial Times, 5/29/81, 9/6/82.

generation computers. The 5-year, £350 million effort would cover four main topics:

1. Software engineering;
2. Intelligent, knowledge-based systems;
3. Interface between humans and machines;
4. Very large scale integration of electronic components.

3. Finally, the British government has repeatedly provided emergency financial assistance to ICL during its periods of severe financial strain. Most recently, this meant a £400 million series of loans over the 1981–83 period. The government has also blocked foreign purchase of ICL on several occasions.

In sum, the expected growth of demand in the U.K. market indicates a modest level of activity through the 1980s; but British computer manufacturers, despite increased public backing, should provide only limited competition (restricted largely to the home market) for their foreign counterparts.

Other Markets

Certain developing countries have added some new reasons and techniques to the traditional motives and methods involved in computer promotion. As through much of Europe and Japan, they see the computer industry as an opportunity for extended growth once a domestic production capability can be established. A pattern of strengthening demand has independently materialized, but it can only be met through unencumbered importation or extensive government involvement. And the vast technological gap between nations like Brazil or Mexico and the United States presents those countries with unusually difficult dilemmas. They must weigh the costs of varying levels of economic inefficiency against their

1. fear of being altogether shut out from the high-tech club;
2. concern over dependence upon foreign sources for computer technologies;
3. need to restore immediately whatever measure of external balance industrial policies can provide.

In the case below the choice has been promotion through elaborate and restrictive computer development programs. For example, performance requirements, trade restrictions, and fiscal and financial assistance have all become central to the high-technology efforts of both Mexico and Brazil. For several reasons—technical constraints, service/support limitations, and growth expectations—public policy has emphasized micro and mini products, leaving the larger mainframes to foreign producers. But the cost to the national governments and domestic consumers of such extensive attempts to restructure their computer market have already proven burdensome, breeding a widespread skepticism as to any eventual closing of the gap between the professed strategy and economic realities. Nevertheless, U.S. firms must still tolerate onerous systems of public control if they are to share in the computer market growth of these newly industrialized countries.

Mexico

In a manner typical of several countries in the developing world, Mexico has singled out the computer industry as a target sector. Its objectives—establishing a sizeable domestic production capability, ensuring that research and development efforts are locally based, and developing the industry's full potential as an exporting sector—inspired a single comprehensive promotion package that now serves as the centerpiece of Mexico's computer policy. This "National Computer Plan", consolidated in December 1981, includes sweeping provisions across the full range of industrial targeting tools. Market access is directly controlled by five basic techniques:

1. A quota system.
2. Imposition of tariffs.
3. An import permit requirement.
4. Selective (i.e., national) purchases by the Government of Mexico and public sector entities.
5. A limitation that foreign firms be only minority partners in joint venture agreements involving local production.

Five additional forms of tax incentives further encourage domestic computer development:

1. A tax credit of 20% of investments in new or expanded production capabilities.

2. A tax credit of 20% of new payroll generated among computer manufacturers.

3. A tax credit of 15% of purchased components manufactured in Mexico.

4. A tax credit of 15% of the purchase price of computer equipment bought from manufacturers registered in the National Computer Plan.

5. Unspecified tax incentives for establishing R&D facilities within Mexico.

In addition, the program provides direct financial encouragement through a hodge-podge of preferential inducements:

1. Special prices for energy (up to a 30% discount off established rates) are made available to computer manufacturers participating in the overall program.

2. Subsidized credit for computer industry development via FOMEX, FONEI, and other federal lending institutions.

3. Government-sponsored (and primarily government-funded) research efforts.

Finally, the plans of foreign subsidiaries are directed towards national goals by the imposition of two statutes over and above the minority ownership provision:

1. Preferential quotas for importing are granted to companies registered under the National Computer Plan, and

2. Export performance requirements are imposed upon investors, ensuring that they "earn" certain minimum levels of foreign exchange.

This elaborate network of industrial policies targeting computer development in Mexico can be expected to interfere with past patterns of U.S. exports to that market. Sales of computer equipment and services in Mexico totalled nearly $700 million in 1981, and the recent 25% per annum growth rate indicated great promise for the foreign manufacturers that dominated the picture. American firms alone controlled some 75% of total shipments (by value), but the combination of Mexico's overall economic crisis and its aggressive program to "domesticate" its computer market could seriously limit the near-term growth potential for U.S.-based computer companies.

The National Computer Plan was only introduced in December 1981, so few conclusions can be drawn at this stage. Indeed, many of the trends it has precipitated (an increase in U.S. shipments of component parts, for example) should moderate the impact of its rather stark objectives and methods. Nevertheless, the National Computer Plan will, over the immediate future, distort Mexico's trade in computer products and services, an area where the U.S. has traditionally played the dominant role. A more extended experience with this targeting package and more detailed study of its apparent consequences should provide important additional clues as to its eventual effects upon both the American and Mexican computer industries.

Keys to the Future
Competitiveness in Computers

The Systems Concept

Perhaps the most striking aspect of future high-technology development will be the continued blending of computers with a broad range of related industries. And as this process advances further, computer manufacturers will find important markets emerging in (and new lessons being learned from) altogether new areas of application. The ability to anticipate and pursue new directions for both hardware and software use will remain a fundamental criterion for corporate success. This wider perspective will necessitate creative appreciation for the potential role of computer technology in systems of a broader nature than simply data processing or information management. The sweeping implications of machine intelligence and advanced automation will influence a full range of economic activities—from farming to high-tech manufacturing itself. And in most of these settings, the effectiveness of particular computer products will depend upon the success with which they have been integrated into the larger systems (whether production, communication, storage, transportation, etc.) at work. The computer company that best adapts its innovation, design, and product to these novel applications will prosper in the marketplace of the future.

Skills

Considerable attention has recently been devoted to the shortage of skills required by the computer industry and related high-technology sectors. Concern has centered around:

1. the declining number of students graduating with an emphasis on engineering, the sciences, and mathematics;

2. the dwindling population of qualified teachers and professors in these critical areas (due largely to disparities in salary between academics and private industry); and,

3. the gradual deterioration of available instruction in these quantitative fields (for reasons of both inadequate staff and aging facilities).

Well-developed human resources have been an important key to American pre-eminence in "knowledge-intensive" industries like the computer field. The importance of a well-trained pool of eventual contributors has led countries as diverse as Singapore, West Germany, and Japan to devote considerable effort to cultivate this resource, for a shortfall of skilled technicians can impose limitations on high-technology development no less serious than financial or production constraints. Several U.S. corporations have already, in recognition of this problem, provided assistance to various universities for purposes ranging from overall technical education to the modernization of laboratory equipment. In some instances, they have even established their own institutes for instruction in such areas as programming and engineering. The options section discusses certain additional possibilities for USG action on the training and education problem.

Research and Development

As always, research and development activity remains a critical ingredient of a competitive computer industry. The "technological acceleration" that has come to characterize most sophisticated sectors ensures that only the innovative survive, and the computer field—at the center of high-technology development—epitomizes this trend.

Recent activity abroad indicates that commercially-oriented R&D expenditures have become an item of priority concern to public officials. As noted earlier, Japan's Ministry of International Trade and Industry has organized several major research projects that involve all of the leading Japanese computer firms. These efforts are designed to address both current industry weaknesses (such as software) and future areas of promise (optical telemetering, 5th generation technology, and supercomputers). This latter group will receive at least Y173.5 billion ($867 million at Y200/$) in direct public funding and Y23.5 billion ($117.5 million) from private sources before their respective conclusions in the mid to late 1980s.

In Europe, much of the public support furnished to the computer industry comes in the form of R&D assistance, and recent patterns indicate that the larger European firms are now concertedly attempting to marshall their forces in areas of long-term study. Phillips and Siemens, for example, have disclosed plans to cooperate in their investigations of sub-micron technology, general microelectronics, computer-aided design, and electronic speech recognition. In addition, as noted above, there are indications of significant cooperation between European governments as they jointly attempt to accelerate their collective computer development.

Within the U.S., computer R&D responsibilities rest primarily with the private sector, a pattern reinforced by recent budgetary trends and changes in tax policy. In the areas of space and defense, real funding levels for R&D have risen from FY1981 to FY1983, but outside of NASA and DOD, most science and technology budgets were trimmed during that same period. This condensation of public activity coincided with passage of the Economic Recovery Tax Act (1981), which included three measures for encouraging private R&D:

1. tax credits for 25% of any increased in corporate R&D expenditures;

2. a two-year suspension of allocation rules governing tax treatment of research and development outlay; and

3. accelerated depreciation for R&D facilities and equipment.

Specific data on the impact of these provisions is as yet unavailable, but despite some conflicting early reports, it is hoped they will catalyze some increase in research and development activity.

At the same time, broader interpretation of antitrust provisions has enabled the formation of selected private-sector R&D consortia, such as the Microelectronic and Computer Technology Corporation (MCC).[16] Led by Control Data Corporation, this group will attempt to exploit the economies of scale and risk-minimization that collective efforts in basic R&D may provide. If this effort establishes a trend, it could mean more frequent inclusion of smaller U.S. companies in the long-term research activities essential to the continued growth and competitiveness of the U.S. computer industry.

Software Capabilities

As noted in the earlier section on software and services, this dynamic aspect of the industry has become an essential component of success in computers. The rising financial and commercial importance of the software field is well-documented. But behind the numbers lies the simple fact that software, often as much as hardware, sells systems. On the one hand, this is an imaginative field where marketable output must not only avoid constraining hardware performance but also open new technological frontiers of its own. On the other hand, it involves more conventional production problems—quality control, standardization, efficiency—that will critically determine the fate of individual firms. Software must continue to embody steady technological improvement while evolving into a mass production commodity. Independent specialists and more diversified packagers will need to move forward on both fronts, and at the same time maintain a creative, far-sighted understanding of future directions in hardware development. On an international level, the growth and prospects in the field have attracted

16. The MCC venture will apparently be constrained as follows: a) it cannot be a profit making enterprise; b) no firm will be allowed more than a 10% interest; c) the Justice Department will also monitor the overall corporate membership, the identity of companies participating in particular projects, and whether or not the risks involved in MCC's efforts are sufficient to justify joint efforts.

the intense interest of foreign firms and governments alike—a sure clue to the mounting challenge that American manufacturers will certainly face in the years ahead. Traditionally, software has been a particular strength of the U.S. computer industry, contributing directly to the sector's prowess at home and abroad. The essential point for the future, therefore, is that only if the U.S. can maintain some of that software leadership will it be able to maintain its overall computer leadership as well.

Foreign Targeting Practices

The computer sector has probably proven the most popular target for industrial policy programs abroad. Because of its critical position at the center of the high-technology field, foreign governments have repeatedly deemed computer development essential to their nations' long-term growth and continued economic well-being. Dramatic improvements in Japan's competitive position across a broad range of industries drew considerable attention to their targeting techniques, and as other advanced nations have since tried to improve their standing in the high-technology race, they have, in some cases, attempted a similar approach, or at least invoked certain similar methods. International agreements have imposed some constraints upon signatories, but financial support, fiscal incentives, and direct public participation in computer development have become commonplace in many foreign markets. These kinds of policies have posed a difficult challenge to both American computer manufacturers and the U.S. government. First, it must be determined the extent to which such practices may or may not erode U.S. competitiveness; and second, it must be decided what type of response, if any, is appropriate. Both are complex questions, on the one hand involving methodological problems, on the other requiring expert analysis of the potential consequences of each available course of action.

Among the developing countries, computer targeting has generally taken a more elaborate, if not more sophisticated, form than the current promotion practices of industrialized countries. In addition to using methods at work in Europe and Japan, they appeal to "infant industry" arguments as justification for imposing performance requirements and establishing direct import barriers for domestic protection. The investment restrictions include export requirements, technology transfers, ownership limitations, and sales ceilings, to name a few. Trade can be controlled through quotas, tariffs, licensing, and national sourcing regulations. The consequences of such measures for the competitiveness of foreign producers are fairly straightforward, and their distorting effects are undisputed. Their limited success in achieving their stated development objectives may eventually lead to more widespread revocation of this restrictive approach. But in the meanwhile, the ability to compete effectively in developing markets in the face of such policy interference has become a vital concern for an American computer industry that is more dependent than ever upon trade and foreign investment for its health and well-being.

Options

A discussion of the Pros and Cons of Proposals for USG Action as Recommended by a Variety of Sources.

Research & Development

A primary issue of great interest is the level of R&D activity in the United States. Several recent policy changes have acknowledged its importance for the nation's future, especially in high-technology areas, but some concern remains over the long-term implications of inadequate R&D expenditures.

The computer industry has always been a leader in terms of its R&D outlays. Revisions in the tax treatment of R&D (embodied in the 1981 Economic Recovery Tax Act) were designed to further stimulate such expenditures. The accelerated depreciation schedules included for R&D equipment should provide significant and secure incentives for corporate investment in this area. The accompanying R&D tax credits, however, may need elaboration to ensure their effectiveness.[17] Insofar as firms' decisions on R&D allocations require long-term planning and more extended lead-times, two years (the applicable period of these current measures) may prove inadequate for generating a broad positive response. A longer-term provision of this type may prove desirable. A second possible shortcoming of the stepwise R&D credit may be its lack of stimulus for the young, fast growing companies that so heavily populate research-intensive sectors, and from which an impressive proportion of technological innovation has emanated. The simple incremental approach embodied in existing legislation may provide the least benefit and incentive to many of those most active in the area of policy concern. Revisions that structure into the formula credits for a baseline, dollar-amount R&D increase (on top of which the 25% schedule would take effect) might somewhat alleviate this problem.

Another set of policy developments in the R&D field involves more open interpretation of anti-trust regulations (see page 41 for discussion). Recognizing that some legitimate economies of scale can be realized through limited inter-firm collaboration in areas of basic research, the Justice Department has given qualified approval to the establishment of the joint venture Microelectronic and Computer Technology Corporation (MCC). Many observers feel that this could represent an important first step towards similar cooperative activity, both in other areas of the computer industry and in other high-technology sectors. But it appears that before any field can reap the full rewards of this new understanding, the ground rules will need to be clarified and secured. The lack of detailed and defensible preconditions will likely deter many valid participants from joining a collective undertaking of this type. At this point, considerable discretionary/interpretive power remains with the Justice Department, the courts' position on such ventures has yet to be clarified, and no protection from civil suits has been provided. In the face of such impediments, legislative action may emerge as the

17. Included are both the "25% of increase" credit and the reallocation of international credits assigned under Code Item 861.

only mechanism able to catalyze full use of this collaborative opportunity.

Personnel

A second, long-term problem for both the computer industry and the economy as a whole is the "skills shortage". Increases in the number of new scientists, mathematicians, and engineers have not kept pace with a growing field's demand for this type of trained personnel. This has in turn led to a depletion of the ranks of qualified instructors remaining in academics. And to complete the cycle, this shrinking number of teachers (in both secondary schools and universities) is less able than ever to educate the larger numbers of trained students needed by high-technology sectors.

One solution to this problem is to provide a greater network of government support for education and training in the areas of concern. Specific recommendations are to:

A. Increase public funding for discretionary improvement and enlargement by educational institutions of programs aimed at the training of scientists, mathematicians, and engineers;

B. Lend more directed public assistance, in such areas as teacher salaries, for the purpose of encouraging qualified academics to remain in their positions as instructors,

C. Provide incentives to the private sector for further corporate training of the necessary personnel,

D. Provide incentives for broader and indirect private sector assistance to educational programs in the maths and sciences (such as the Computer Equipment Contribution Act considered by the Congress in 1982).

These kinds of measures would both facilitate scientific education and help pique the interest of greater numbers of prospective students in the designated fields. It could also assist in retraining of workers whose skills have become obsolete because of shifts in the U.S. production base.

However, special caution would need to be exercised to avoid intensifying the current competition between industry and academia for skilled people (see, for example, the proposed salary assistance for teachers). Also, increased budget support would be required under any of these programs unless current resources were transferred from the liberal arts to the sciences, a move that could generate considerable opposition from other affected interest groups.

Another solution is to leave the necessary adjustments to the marketplace. For some, this may offer a more efficient alternative to the kinds of government involvement implied in the policy options outlined above.

But, the adjustments needed to restore equilibrium between the supply and demand of technical skills under a laissez-faire approach may require an inordinate period to complete. The shortage of qualified personnel is an immediate problem which, if not addressed soon, could have adverse long-term consequences for the American economy. In other words, the employment market may function inefficiently in translating sudden changes in demand through educational institutions into shifts in the training and eventual supply of properly equipped graduates.

Countering Foreign Competition

A third area of concern for many involved in the computer industry is the proliferation of foreign government programs aimed at the development of a domestic computer capability. These may affect the competitive position of U.S. manufacturers.

Adopting comparable targeting practices could spur development in any of several areas of the computer industry. Comparable targeting practices could provide additional demand stimuli, encourage risk-taking among producers, avoid undesirable waste and duplication in certain R&D areas, and presumably, place U.S. manufacturers on an "equal footing" with their foreign counterparts.

Certain minor types of support (especially basic research in government and university laboratories) and special tax credits for R&D are already in place. More liberal interpretation of anti-trust regulations has also enabled some joint research efforts to be organized between computer companies. More sweeping measures would require a fundamental change in the current philosophy of business-government relations in the U.S. Such revisions could also either a) shift competition in computers from production programs to support programs as other countries in turn attempt to provide the most generous terms for development or b) precipitate the introduction of less palatable trade barriers, such as tariffs, quotas, etc. And again, industry-specific USG policies would invoke demands for equal treatment across a whole range of American sectors that feel similarly victimized.

Protecting the U.S. Market could result in an eventual dismantling of selected foreign industrial policy programs if this is accepted as the price for regaining access to the U.S. market. However, the vagueness of many foreign provisions and the inherent competitiveness of the industry may make injury to U.S. producers impossible to prove under the GATT. And since many imports will enter under agreements involving American firms, such action could even have ambiguous short-term consequences for U.S. interests. This solution also fails to address the question of third markets—and protective diversion of foreign exports could erode U.S. market share abroad. Finally, such unilateral action raises the spectre of a full trade war, an eventuality that could seriously damage the U.S. economy across a much broader range of products and industries.

A vigorous U.S. program to *counter targeting programs* through strict enforcement of U.S. trade laws (under Section 301) could produce case-by-case agreements as to how an equitable trading environment could be restored. Historically, this has been a successful process, with only rare invocation of Executive Authority to impose unilaterally reciprocal restrictions. Above all, active enforcement would lend integrity to the legal structure now in place.

Presently, most U.S. trade laws only emphasize temporary relief and adjustment assistance where damage is found, offering little to actually discourage targeting practices. The resources required of firms to pursue trade action cases, and the often lengthy period between violation and judgment, may discourage many (particularly smaller companies) from invoking what provisions are available. While several complaints are still in decision at this time, Section 301 (the mechanism relevant for most targeting problems) would also appear not to deal with the question of third markets, and many nations may well decide that the benefits of promotion policies still outweigh the costs of American enforcement. Finally, there remains some question as to the GATT-legality of certain responses of this type.

Negotiation through bilateral channels for country by country *removal of the most restrictive practices* (such as performance requirements and blatantly protectionist trade barriers) could give rise to a consistent and principled U.S. strategy for dealing with this type of restriction in the context of particular bilateral relationships. The U.S. approach to discussions of this kind would have to take into full consideration the range of other economic, political, and security interests at stake. Because these may vary greatly from case to case, it could prove difficult to develop any set of policy positions on targeting in developing countries that appears coherent and non-arbitrary.

Export Controls

A final concern for industry and government alike has been the effect of export controls (whether COCOM or unilateral restrictions) upon computer sales and computer firms.

USG policies in this area will clearly be attempts to balance sometimes conflicting objectives, and any effort to dictate a binding solution in this context would prove highly problematic. However, the upcoming renewal of the Export Administration Act does provide an opportunity for debate over the methods and content of export controls. As part of the review, it could be appropriate to:

A. re-emphasize the priority of technology transfers over product transfers as a guiding principle for security concerns,

B. underscore the broad damage to commercial interests that results from perpetuating the United States' reputation as an "unreliable supplier".

C. highlight the fact that export markets in high-technology fields also contribute to national security by expanding the U.S. military/industrial base.

D. urge all policy-makers involved to give full attention to the competitive interest of the relevant manufacturers, noting in particular the disproportionate burden that export restrictions can place on smaller computer firms, and

E. formalize this advocacy role by involving Commerce Department industry and trade policy specialists in future discussions of computer trade controls.

Computer Industry Overview

BY VICO E. HENRIQUES
PRESIDENT
CBEMA

I have tried to collect ideas and data that will portray the current state of our industry, both domestically and internationally, and the key facts of life which determine our member companies' international competitiveness.

The computer industry is at the heart of a growing collection of intertwined but separate industries which depend critically upon the use of digital technology. While the inherent advantages of a reprogrammable control unit have caused computers to be at the heart of devices as different as space shuttles and automobiles, the industry we are talking about here is that of commercial off-the-shelf computers and business equipment which are found in information systems and networks. This includes computing equipment from the

81

home personal computer to the supercomputers NOAA uses for weather forecasting. It includes the computer software provided with or purchased for use on these computers. It includes all of the supporting products such as terminals, memory and printing equipment. Last but very importantly, it includes the support and maintenance services which keep the equipment and software running.

The computer industry began only around 1950, yet today is one of the United States' major industries, and even more critically, is viewed by most industrially-oriented nations as the key industry for their economic future.

The following four tables give some measure of the amounts of business and the rates of growth being experienced by our industry. Let me summarize these briefly by saying that in 1983, worldwide industry equipment and software revenues, excluding on-line services and user programming investments, will amount to an estimated $92.5 billion dollars.

Table 14
Worldwide
Computer and Business Equipment Industry Revenues 1965–82

Description	1982E	1981	1980	1975	1970	1965
Total Revenues	113.300	101.600	90.600	46.200	20.500	4.500
Total DP Equipment Revenues ($Bil)	69.000	61.800	55.100	27.900	10.500	2.400
Application of Software Revenues Excluding Contract Program ($Bil)	23.500	21.000	18.000	6.500	2.500	.200
Total BE Equipment Revenues ($Bil)	16.500	14.700	13.500	9.300	6.000	3.500
Business Form Revenues ($Bil)	4.300	4.100	4.000	2.500	1.500	.900

Table 15
CBEMA Members' Gross Revenue Distribution

	Foreign	Domestic
1981	37%	63%
1980	37%	63%
1975	40%	60%
1970	35%	65%
1965	25%	75%
1960	18%	82%

In addition, we estimate $20 billion dollars in business equipment and business forms revenues, much of which either supports or utilizes the application of computers in making business more effective. The dynamics of the industry are supported by heavy R&D investments. In 1981 our CBEMA members invested $3.855 billion in R&D, a 32% increase over 1980.

International trade is important to our industry. In 1981, the domestic revenues of our member companies were 63 percent of their world-wide revenues. That means that 37 percent of our member companies' revenues come from international operations. In 1982,

Table 16
CBEMA Members' R&D Expenditures (Billions of $)

1982	4.700E
1981	3.855
1980	2.915
1975	1.660
1970	.921
1965	.172
1960	.050

Table 17
CBEMA Member Company Profile—Employment 1960–1982

Description	1981	1980	1975	1970	1965	1960
Total Industry	1,550,000	1,530,000	1,240,000	1,143,000	630,000	365,000
Employment Domestic Industry	1,160,000	1,145,000	1,000,000	900,000	600,000	300,000
Employment Foreign Industry	740,000	720,000	600,000	585,000	450,000	246,000
Employment	420,000	425,000	400,000	315,000	150,000	54,000

total computer industry exports from the U.S. were $8.88 billion and imports were $2.14 billion, giving us a favorable trade balance of $6.74 billion. In addition to this trade balance, we must add a large flow of revenues from licenses, royalties, dividends and other "invisibles." The amount of this flow is difficult to quantify from the data which we have, but it constitutes approximately 10 percent of international revenues, or about three billion dollars, and is growing. These data point out that this industry is truly multinational. We do not, and cannot because of the kind of business we are in, export solely from the United States. On the other hand, these data point out that almost all computer companies, even the very smallest, quickly engage in international trade, and that international trade remains a significant part of their business.

As a company's activity expands abroad, it soon requires the establishment of a local presence beyond that of a distributor. Frequently, branches and then full-fledged subsidiaries are established. Because of the requirements of the local markets, establishment of manufacturing operations which cater to the standards and requirements of those markets are likely to follow after obtaining a significant amount of business. Ultimately, development and even research operations may be established in major areas.

Development activities are frequently necessary to support local manufacturing activities when the requirements for the marketplace are significantly different from those in the U.S. A good example would be to optimize products for sale within the European economic community. Development and research activities also are established abroad because that is where the expertise may be. Despite the intense concentration on research and development within our industry domestically, no environment would be more productive for research and development into the input and output of non-roman alphabet characters than Japan and China, simply because of the deeply-felt need for continuing to use the ideoform writing, leading to a great deal of attention on developing devices that can handle it. Thus, we see a two-way flow of technology within companies which are established abroad that integrates with the worldwide marketing and manufacturing operations to support the global marketing approach required in the industry.

Many computer products require a market greater in size than the U.S. market, and for some products, only a world-scale approach will support the investment necessary for the product.

The industry's products are non-sectoral in that they are used by practically every business, they are used by governments, and with the advent of personal computers they are now used by the individual.

This is perhaps a unique occurrence in industrial development. The implications have been that the industry has expanded greatly over the last three decades, and continues to expand, even in the current recession, because of the potential for our products to enable others to increase their productivity and reduce their costs. There are steps which the U.S. government can take, some in traditional trade policies and some in domestic policies which would enhance our international competitiveness in the face of the concerted activities of our trading partners. However, we do feel, overall, that the opportunities for the United States computer industry are excellent.

The Computer Industry: Restrictions and Performance Requirements

BY EDSON DECASTRO
DATA GENERAL CORPORATION

Access to World Markets

The computer industry continues to encounter investment restrictions and performance requirements that deny or impede access to important world markets. Brazil, Mexico, South Korea are the most current examples. Details of these barriers are well known. The trend is rapidly spreading.

Mexico

Last year Mexico introduced an integration plan through which they hoped to establish a domestic computer industry. Details of this plan are now being changed, due to the Mexican economic situation. Regardless, companies like Data General are severely limited in the number and value of import licenses into Mexico. And we are not allowed to establish there unless we agree to surrender controlling interest, transfer technology and build substantial manufacturing facilities.

Brazil

Since 1978 Brazil has been trying to establish a visible domestic computer industry through government protection and subsidy. Over the five years since then, market access has been granted to only those who agree to transfer of technology. There have been few takers and import licenses for shipments to Brazil are few and far between. Their effort to build a domestic industry continues to be heavily subsidized by the Brazilian government and is notable for its lack of success. Realizing their inability to produce newer 32-bit computers, they are once again, as in 1978, trying to strike bargains with American firms. The deal is more or less the same: Access to the Brazilian market in return for technology transfer.

South Korea

South Korea's markets have been alternatively closed and opened to us. At present, foreign investment laws are being used to fashion a new computer integration plan that threatens to once again restrict our ability to sell products there.

It is an approach to imposing non-tariff trade barriers for the protection of domestic industries that began in the non-industrialized world and now threatens to spread to industrialized nations (Canada is an example). It is particularly appealing during periods of economic decline.

U.S. Industry Structure and It's Future

Investment restrictions and performance requirements threaten the structure and the future of the U.S. computer industry.

COMPUTER

U.S. computer companies are reliant on international business and derive a substantial portion of revenues from exports. Because of the rapid pace of technological development, the industry is capital intensive. Growth and development rely heavily on an expanding revenue base. This can only come from full participation in established and developing global markets. Reliance upon domestic markets alone is not enough.

Companies need to be close to their customers in order to adapt their products to local/regional application needs.

The sale of computer equipment requires an ongoing commitment to service and support of products once sold. Companies need access to foreign markets in order to meet these commitments to their customers.

There is no well-enunciated U.S. policy on investment restrictions and performance requirements. There is a lack of coordination between government agencies on such matters. There are not adequate resources within the U.S. government to deal with this issue. Our industry needs all of these.

Since there is no multilateral vehicle for addressing investment restrictions and performance requirements, alternatives must be found.

Last fall's GATT ministerial meeting suggests that the GATT is unable, for the time being, to deal with the issue. In the short term, bilateral solutions are the only viable alternative.

Five possible solutions are:

1. Many of those non-industrialized nations with firmly-established restrictions on foreign investment and performance requirements are experiencing financial crisis. Some teeter on the brink of bankruptcy. They turn to the U.S., the World Bank and the International Monetary Fund for help in the form of new credits and renegotiation of debt repayment. U.S. assistance, either direct or indirect, should be predicated on the relaxation of restriction.

2. The U.S. should adopt a "no nonsense" policy of opposition to such barriers to free and fair trade and to national industrial policies that prevent accordance of national treatment.

3. Based on such a policy, the efforts of all U.S. government agencies should be well-coordinated to reach these objectives.

4. Greater resources should be devoted to bilateral negotiations of national treatment accords, starting first with those nations that represent the most substantial markets.

5. U.S. trade and foreign assistance programs should be used as incentives for other nations to relax restrictions.

U.S. Response to Industry Targeting Practices

BY STEPHEN G. JERRITTS
SENIOR VICE PRESIDENT
HONEYWELL, INC.

I would like to express my appreciation to the Secretary of Commerce for the work which his department and others in this administration are doing to promote international trade. This is a subject of tremendous importance to Honeywell, and one to which we feel an increased need to devote our attention.

What I'm going to say here makes a case *against* protectionism and *for* a more competitive America. There can be no denying that some countries have adopted mechanisms to aid their domestic computer industries. And there can be no denying that the American people feel a growing frustration with the economic relations between this country and others, particularly Japan.

Honeywell believes that many of Japan's protectionist barriers have come down, and that more will come down with a continued aggressive negotiating stance by the Administration. But elimination of these barriers alone is not enough to help keep U.S. industry competitive, either here or in foreign markets.

There are four key elements in which we believe government support can help the private sector become more competitive. The four areas are: *foreign trade barriers; fiscal policy; government supported research and development;* and *industry cooperation.* Export control policy is also a critical element but that will be discussed next by John Lacey of Control Data.

I would also like to point out that we do not believe that the U.S. should adopt the industry targeting methods used by other countries. Each country has to develop policies that reflect its own culture, its own history, its own economy. We believe that the U.S.' economic strength is its free market system. We need policies that improve the efficiency and adaptability of that system.

Foreign Trade Barriers

The United States now has on the books a variety of measures to deal with unfair foreign trade practices.

Remedies provided in these laws include import restrictions, countervailing and anti-dumping duties, vigorous enforcement of patents and copyrights, and enforcement of rights under trade agreements including GATT. We are pleased that both the U.S. Trade Representative's Office and the International Trade Commission are showing a willingness to take aggressive action with these existing laws.

Unfortunately, our complex and open legal system may result in the remedy's coming long after the damaging fact. The 1979 Trade Act sought to speed up the fact-finding and ruling process, but much more needs to be done.

Much more also needs to be done within GATT. GATT provisions need to be extended to services and investment, and negotiations must also begin on the treatment of high technology industries.

Very little was accomplished in these areas at the recent GATT Ministerial but the U.S. should continue to pursue them.

As proposed in the Senate's Reciprocal Trade and Investment bill last year, the U.S. Trade Representative's Office needs to have its responsibilities extended to include monitoring as well as negotiating. In particular, the Trade Representative's office needs to search out and publicize countries and regulations that treat U.S. and other foreign firms differently from domestic firms.

Congress should also restore the President's authority to negotiate tariff reductions in specific industries.

Fiscal Policy

We believe the recent steps to increase cash flow in some industries by speeding depreciation schedules, to stimulate investment in new businesses by reducing capital gains taxes, and to lower income taxes are very positive. Over a period of time these steps will increase the productivity and competitiveness of American industry. It is true that the 1981 and 1982 depreciation changes have had little effect on cash flow in the computer industry. However, we do believe they have had a positive impact on the economy as a whole, and, not incidentally, on many of our important customers.

But there are some things which the government could do in the fiscal area that would be of great benefit to the international competitiveness of the computer and other high technology industries:

Research and Development Credit

The 25 percent credit on increases in qualifying R&D expenses needs to be made permanent and to be liberalized. In fact, we believe that it would be a positive step to make all R&D, not just the increment, qualify for the credit. Although the credit has been in place only a short time, and many companies are only now beginning to take it into account when determining their research programs, we believe it is having a positive effect. Unfortunately, the

93

credit is due to expire after 1985. From now until 1985 is not a long enough period of time to expect a wholesale improvement in the R&D plans of American industry. Industry action is also being held back since early termination of the credit seems to be turning up on many lists of possible "revenue enhancers."

Section 861

The present two-year suspension of the Treasury Department's regulations on the allocation of R&D expenses also needs to be made permanent. The Treasury regulations would have limited the deductibility of R&D expenses for companies who sell their products overseas.

DISC

DISC has provided cash-flow benefits for U.S. exporters by deferring some federal income tax on export sales. It has been attacked by other countries as a violation of GATT rules. If DISC must be replaced, an alternative must be found that is consistent with GATT rules and that provides equivalent cash flow benefits.

Federal Deficit

We are not here to discuss the specifics of reducing the federal deficit. But it must be reduced if we are not to have higher interest rates and a resulting early return of inflation.

Government-Supported Research

The federal government has played a major role in industrial development, with substantial fall-out from government contracts benefiting related products subsequently sold in the private sector. The computer industry, in fact, grew out of government contracts in the 1940s.

At the present time, the Defense Science Board has rated the

VHSIC program as the number one DOD technology program. This program, which was initiated to jump the "state of the art" of military integrated circuits, is already having an impact on the integrated circuit industry. It stimulated the industry to accelerate programs in finer-geometry integrated circuits. One criticism of the VHSIC contracts is that they were several years too late.

More federal funding of VHSIC-type programs in basic technologies affecting the future of our electronics industry will be essential to stay ahead in those technologies that are too far ahead of commercial feasibility to attract commercial funding at adequate levels.

Nearly six percent of the federal budget is for research and development, primarily through DOD, NIH, NASA, DOE, and the National Science Foundation. The Reagan Administration's proposal to increase funding to universities through the NSF is a good sign, as that budget has seen no real growth over the past four years. There also needs to be some discussion of setting overall goals and strategies for the government's investment in R&D, rather than negotiating it on an annual basis through countless budget line items.

Industry Cooperation

Of all the four areas I have mentioned, we believe that this is probably the most important. Individual U.S. corporations face many research and development projects too large and costly for them to handle individually. But they are reluctant to engage in cooperative efforts with other companies because of anti-trust risk. There are some innovative efforts in the cooperative R&D area which minimize these risks.

One example would be the Semiconductor Research Consortium created by the Semiconductor Industries Association. The approximately 30 members of this consortium contribute on the basis of their integrated circuit use and/or sales. The consortium distributes its funds in response to proposals from universities and will initiate new areas of research most needed by U.S. industry.

Another example is the Microelectronic and Computer Technology Corporation established under the leadership of Control Data.

MCC is similar to a limited partnership for R&D in which companies combine to form a vehicle that performs the R&D at arm's length and licenses the results to avoid anti-trust problems.

Some reduction of the anti-trust risks in cooperative R&D came with the publication of the Justice Department's liberalized guidelines in 1980 recommending the issuance of "business letters" to specific joint ventures. These letters indicate that the Justice Department has no intention to attack the joint venture on anti-trust grounds. But the letters are no assurance that Justice will not change its mind in the future, nor are they any protection against private anti-trust suits.

The law needs to be modified to grant binding anti-trust exemption, applicable to private as well as government suits, for approved R&D joint ventures. Such exemptions would enable U.S. companies to combine complementary resources and skills to accelerate innovation and avoid duplication of effort.

Cooperative efforts may also be needed in areas other than R&D. The recently enacted Export Trading Company Act, for example, recognizes the need for marketing cooperation in export sales. Another area for consideration might be certain declining U.S. industries. This could be achieved through an official identification of these industries, and the establishment of definite anti-trust exemptions for the appropriate mergers, consolidations, transfers, rationalization and other cooperative action between companies within the industries.

In summary, we need to keep negotiating pressure on Japan and other countries to eliminate the remaining barriers to trade and investment. But other advanced industrial countries will always be formidable competitors for many of the products we make. These steps I have outlined will, we believe, enable us to help ourselves maintain and improve the competitive position of the United States.

U.S. Controls on International Trade

BY JOHN W. LACEY
EXECUTIVE VICE PRESIDENT
TECHNOLOGY AND PLANNING
CONTROL DATA CORPORATION

Control Data strongly supports the effective administration of the current Export Administration Act for export promotion and its enforcement for national security purposes.

We are very concerned, however, over the West-West export controls which are in place and which are becoming more extensive as a result of the efforts by the U.S. Government to deny or delay the acquisition of products or technologies by our adversary nations.

Our industry has a good record and a deep interest in protecting its proprietary technology. This industry is multi-national in nature. The United States is no longer, if it ever really was, the only source of technology in our industry. There must be free flow, both ways, of research and development results within the corporate structure if we are to remain competitors in the international marketplace.

Control Data also questions those who have suggested that restrictions be placed on the results of basic research which are in a real sense "vital" to the competitiveness of U.S. industry. We believe the free exchange of such research, excluding classified work carried out for national security purposes, benefits the United States and its allies. The exchange of ideas is the basis for much of our industry's development work.

A technological parity exists with our trading partners. With the realization of that fact we must develop policies toward effective controls which: (1) will retard the flow of technology to our adversaries, (2) recognizes our industry's multi-national nature and (3) does not further, but removes, the causes of our growing reputation for being unreliable suppliers.

The remainder of this paper will address eleven specific areas of the export controls with a view toward possible future implementation strategies.

The Need for U.S. Industry to Export

There is an obvious need for a positive balance of trade and for a strong United States economy. It must be understood that many technologies developed within the United States for commercial use are more advanced than the technologies currently used for military purposes. The Department of Defense must rely increasingly on technologies developed for commercial purposes. The quality of commercial technology available to the Department of Defense from U.S. companies depends upon the competitiveness of those companies in the world market. Therefore, export control measures which put U.S. high technology companies at a competitive disadvantage to foreign companies in Western Markets adversely affect our national security in the long run and must be avoided to the maximum extent possible.

Multilateral Controls

It is essential to establish effective export controls on goods and technologies which can make a direct and significant contribution to the military capabilities of specific adversary countries. Such militarily critical goods and technologies can be acquired in many western industrial countries. Therefore, multilateral agreement with our allies and other non-adversary countries is the only effective means to deny access to such goods and technologies by adversary countries.

Foreign Policy Controls

The United States Government is one of the few governments in the world that imposes foreign policy controls on its exporters. An objective examination of the effectiveness of foreign policy controls which have been implemented over the past years shows that these controls have not served their intended purpose. Indeed, in most cases, the opposite effect has been achieved.

The imposition of foreign policy controls by the President is one of the major reasons why U.S. exporters have gained the reputation over the past years as being unreliable. With increasing regularity, this is causing foreign customers to no longer consider U.S. exporters for their procurements. This coupled with the increasing availability of equivalent commodities from foreign manufacturers is causing U.S. industry to lose increasing segments of the foreign market.

The imposition of export controls for national security reasons gives the administration plenty of tools to protect the national security.

It seems that the imposition of foreign policy controls has been for the primary reason of the Reagan administration's desire to send signals to certain countries with which it is not pleased. For the U.S. exporter the result is a spigot-like mechanism being turned on and off, causing industry to be regarded as an unreliable supplier along with severe economic losses that are associated with this on-off process and of course the jobs that are lost in the process.

In summary, the ability of the President to impose foreign policy controls should be severely curtailed if not removed completely. The national security would not suffer because of this because of the very adequate control mechanisms available to the President under the national security controls process.

Serious consideration should be given to trading freely with all countries with whom the United States has diplomatic relations within the constraints of national security controls. This is the process by which most other countries operate and they do it quite successfully recognizing the need for a strong internal economy and the need to not present their exporters with the disincentives associated with the on-off spigot of foreign policy controls.

Exports to COCOM Countries

The requirement for individual validated licenses for exports of goods or technologies subject to multilateral controls to COCOM countries should be removed.

The purpose of controls is to deny adversary countries (not allies) access to specified goods and technology. This purpose can be achieved effectively only by multilateral controls. Individual review of license requests for exports of goods or technologies subject to multilateral controls to allies is irrelevant to the purpose of the controls and diverts enforcement resources from achieving that purpose.

Reexport Controls Within COCOM Countries

The requirement for reexport controls on goods and technologies exported to COCOM countries when such goods and technologies are subject to multilateral controls should be removed.

Controls will be effective only if all sources, not only the United States, are controlled. As a practical matter, this can only be achieved through multilateral controls. Under such controls each COCOM country must be responsible for controlling reexports from its territory to adversary countries.

Exports to Third World Countries

Less stringent licensing requirements should be imposed on exports of goods and technologies to non-adversary, non-COCOM countries which agree bilaterally with the United States Government to impose controls on exports and reexports which are similar or identical to COCOM controls.

Because all sources must be controlled, every effort must be made to expand multilateral controls to as many countries as possible. Effective bilateral agreements with non-adversary neutrals would allow allocation of enforcement resources to more serious problems such as illegal acquisition efforts by adversary countries. The precise level of controls under a bilateral agreement would depend on the stringency of the agreement.

Foreign Availability

What industry needs in the area of foreign availability is an implementation of the spirit of the words covering foreign availability in the Export Administration Act of 1979. There needs to be a thorough assessment capability for foreign availability within the Department of Commerce and associated export control agencies. This assessment capability must be extremely responsive to the needs of industry in a situation where a United States company is competing for business with a foreign manufacturer who will not experience any licensing delays. The United States company must have a very quick assessment of foreign availability such that it can continue to compete for that specific business.

Delays in foreign availability assessment and verification will serve only to deprive the United States manufacturer of the ultimate contract because the foreign competitor has in the meantime already delivered the commodities and taken the business away from the United States manufacturer.

In fact, U.S. business has been experiencing this scenario more and more over recent years. There does not exist at this point in time a viable foreign availability assessment capability within the Department of Commerce much less the capability to respond to the needs of U.S. industry within any reasonable timeframe. There also does not exist within the United States Government a capability to verify claims of foreign availability that may be submitted from manufacturers as part of their export license applications.

In summary, there needs to be a viable foreign availability assessment capability that is responsive to the needs of the U.S. industry within a very short timeframe so that U.S. industry can remain competitive in the foreign marketplace.

Indexing

As time passes the state of the art in technology sophistication and performance increases. The current export regulations, both within the U.S. Government and within COCOM take into account the technology levels that existed prior to 1974. U.S. manufacturers have long since gone beyond the technology levels that existed in 1974 in their current product lines.

U.S. manufacturers cannot afford to continue to build obsolete product lines for export which would fit under the current export control guidelines when our foreign competitors are building products and technology which are state of the art and contemporary. Consequently, U.S. manufacturers are finding it more and more difficult and sometimes impossible to export contemporary products. Meanwhile, foreign sources of the same and equivalent products are able to take the business away from U.S. manufacturers because the export controls which govern their exports are either non-existent or less stringent than those imposed unilaterally on U.S. exporters.

This situation must be rectified so that the indexing of the export control guidelines is kept up with the advancement that occurs over a period of time in technologies and products so that the United States can remain competitive in the foreign market place and is not unnecessarily impeded by a lack of an indexing process within the U.S. export control guidelines. The Export Administration Act of 1979 clearly calls for a periodic indexing of the guidelines for export controls. This has not happened.

Unilateral Controls

Export controls imposed unilaterally by the U.S. Government on goods and technologies should terminate one year after the date on which they were imposed. If COCOM countries agree to multilateral controls on such goods and technologies, an extension or renewal of such unilateral controls should be prohibited.

Militarily Critical
Technologies List

The contents of the Militarily Critical Technologies List must not be unilaterally imposed on United States exporters. The proper way for the MCTL to be implemented is to first seek multilateral agreement on those technologies and products which must be controlled from export to our adversaries and then, and only then, should the regulatory language which encompasses the contents of the MCTL find its way to the United States Commodity Control List.

Only multilateral controls will be effective and an imposition of unilateral controls on U.S. manufacturers will only serve to decrease the market share in the foreign market place of U.S. exporters.

Enforcement

Government enforcement resources must be focused effectively on illegal activities. This goal can only be achieved through measures which promote voluntary compliance by responsible companies to the maximum extent possible. This in effect places export controls on free world trade into the framework of "pre-emptive" controls whose main purpose is not the regulation of U.S. business but the forcing of Soviet acquisition programs into the open.

Effective regulation under the export controls must force adversary countries to employ illegal means if they attempt to acquire militarily critical goods or technology from the West.

103

Summary of Specific Issues Raised by the Computer Industry

Summarizing the current issues of particular concern to the computer industry:

■ The industry requires both the domestic and international markets in order to sustain its characteristically high level of investment in R&D.

■ Investment restrictions and performance requirements increasingly imposed by developing countries are serious protectionist barriers which could cause long-term injury to the computer industry. There should be a well-enunciated U.S. policy against investment restrictions and performance requirements.

■ Foreign industry targeting policies have an adverse effect on U.S. commercial interests. The industry does not believe that the

U.S. should adopt the industry targeting methods used by other countries. Rather, we need to address the competitive problem through domestic policies that improve the efficiency of the U.S. free market system, including the following:

a. Further removal of foreign trade barriers through bilateral and multilateral negotiations and stronger GATT enforcement mechanisms.

b. Strengthening tax provisions that affect research and development and U.S. exports, including permanent R&D credits, deletion of IRS Reg. 861-8, and a DISC substitute.

c. Development of long-term U.S. policy towards non-defense research to promote technological innovation in the commercial marketplace. The defense market is too small to provide adequate stimulus to widespread technological advance.

d. Removal of outdated restrictions on inter-corporate cooperation in R&D.

■ Stronger multilateral mechanisms for denying militarily significant products and technologies to the USSR and other designated adversary countries are needed, while reducing product and technology controls on trade with countries agreeing to support similar controls on trade with adversaries.

■ The U.S. should resist the temptation to apply export controls for reasons of foreign policy since history shows that such controls do not work when applied only by the U.S.

3
The Robotics Industry

A Study of the Competitive Prospects of the U.S. Robotics Industry

BY ROBERT ECKELMANN
U.S. DEPARTMENT OF COMMERCE

Purpose and Summary

This paper is designed to:

1. assess the international competitive position of the U.S. robotics industry;

2. pinpoint the major foreign and domestic challenges to American robot producers; and

3. suggest possible options for USG policies affecting the sector's international standing.

The U.S. robotics industry demonstrates the vitality and dynamism typical of a rapidly growing high-technology area. Sales and production are expanding at a vigorous 35-40% per annum rate, and most forecasters expect that trend to continue through 1990.

However, the case of robotics remains somewhat of an exception among high-technology sectors for one important reason—the U.S.

is not the world leader in all but a few, strictly technological aspects of the industry. Instead, both political and economic factors, including a comprehensive industrial policy program aimed at robotics development, have spurred the Japanese to a position of clear preeminence. Output and consumption in Japan have stabilized at 3-4 times U.S. levels, and that ratio should persist well into the mid-1980's, enabling them to maintain their majority share of the world robotics market through most of the decade. A select group of European nations could also play a significant role in the most advanced, research-intensive areas of a highly innovative field.

Unfortunately, studies of the robotics industry remain plagued by problems of data availability and interpretation. No single definition of a robot is universally accepted. The lack of a classification system that organizes robotics trade into coherent categories further complicates international comparisons. And an intensely competitive atmosphere discourages firms from disclosing the additional information necessary for thorough analysis.

Nevertheless, important conclusions can be drawn. Robot capabilities will steadily increase and diversify, and those improving qualities will enable the gradual incorporation of robots into Flexible Manufacturing Systems (FMS) that combine various forms of factory automation into fully mechanized and self-sufficient production processes. The robotics industry itself will certainly continue to boom over the next 10 years, with American firms sharing in much of that growth and contributing heavily to the sector's rapid technological progress. But during the 1980's, Japanese dominance will remain the general rule, with selected European firms providing additional competition in the more sophisticated end of the industry.

The profile closes with an options section that discusses possible USG action in three policy-sensitive areas of robotics development:

1. the problem of sluggish domestic demand for industrial robots,

2. the U.S. response to foreign targeting practices that promote overseas competitors in the robotics field, and

3. the standards/regulatory aspect of robot production and usage.

The following brief examination of these issues can hopefully serve as a launching point for further consideration of the options available to both government and industry in dealing with the challenges facing this U.S. industry.

Definitional Problems:
The Product and the Sector

No single definition of a robot is universally accepted; as a result no single description of a "robotics industry" has proven entirely satisfactory. But the following two sections, by summarizing the current debate and offering descriptive information, should provide the basic perspective necessary to analyze and understand this rapidly forming sector.

A Robot

Both the Robot Institute of America (RIA) and the Japanese Industrial Robot Association (JIRA) have attempted to resolve the question of product definition with the following proposals:

RIA—A robot is a reprogrammable multifunctional manipulator designed to move material, parts, tools, or specialized devices through variable programmed motions for the performance of a variety of tasks.

JIRA—A robot is a machine 1) capable of performing versatile movements resembling those of the upper limbs of a human (arms/hands) or 2) having sensing and recognition capacity and being capable of controlling its own behavior (intelligent robot).

Despite considerable disparity between these statements, they are not irreconcilable. The narrower RIA criteria generally result in the following robot types.
1. Programmable, servo-controlled, continuous path;
2. Programmable, servo-controlled, point-to-point;

3. Programmable, non-servo robots for general purposes;

4. Programmable, non-servo robots for die casting and molding.

The less restrictive Japanese category simply includes one additional major classification, "mechanical transfer devices" (MTDs), and a few minor categories of fixed sequence machines. But unfortunately, inconclusive progress has been made in translating this rough conceptual correspondence into truly compatible international statistics on robotics.[1] Analysts generally prefer the RIA proposal, which, based on its endorsement at the 11th International Symposium on Industrial Robots (Tokyo, October 1981), will be applied throughout the remainder of this report. However, its use nevertheless requires that subsequent conclusions often be based on estimates alone.

Still, regardless of semantic quibbles that will certainly be resolved as the industry matures, there exists a clear line of products called "robots." And, since the language involved in the above standardization attempts remains technical and vague, a brief survey of the basic features, functions, and price categories of industrial robots (IRs) should provide a clearer sense of the terminology in question.

The Make-Up of an Industrial Robot (IR)

1. Its Mechanical System—grippers, robotic "hands", or similar special-purpose devices (welding or painting mechanisms, for example);

2. Its Measuring or Servo-System—precisely controls and positions the "arms" of the robot.

3. Its Computer-Control System (not relevant for MTDs)—contains specific programming tasks and sequences that direct the robot.

---

1. Recognizing this problem, the Commerce Department's Bureau of Industrial Economics is now working towards its resolution with the OECD and the Working Party on Engineering Industries and Automation, ECE.

Robotics

The Principal Uses of Industrial Robots (IRs)
1. Materials Handling
2. Machine Loading
3. Spot Welding and Arc Welding
4. Stamping
5. Casting and Forging
6. Spraying and Finishing
7. Assembly

General Price Categories for Industrial Robots

Materials Handling	under $20,000 unit
Welding, spraying, drilling, and similar single-purpose IRs	$20,000 to $50,000/unit
Multifunctional	over $50,000/unit

The Robotics Industry

The Producers
Over 50 U.S. firms currently manufacture industrial robots (IRs), and the number grows month-by-month. The following table lists the major U.S. producers, with estimates of their robot sales and respective shares of the U.S. market.

These robotics companies fall into three general categories:

A. established robot producers that either began in the field (Unimation) or entered early based on their machine tool/processing system emphasis (Cincinnati Milacron, Prab Robots, etc.);

B. venture capital groups, typically smaller operations spurred by innovation and the prospect of growth (Automatrix Inc., Advanced Robotics Corp., Mobot, etc.);

C. major corporations (GE, IBM, Westinghouse) with an existing high-technology emphasis seeking both to parlay their related strengths into a share of the robotics boom and to support, through robotics development, their other interests in factory automation.

Table 18
U.S. Robot Producers: Domestic Sales (mil $) and Market Shares (%)

	1980	1981	1982*	1983*
Unimation[1]	40.0 (44.4)	68.0 (43.8)	65.0 (32.1)	60.0 (22.8)
Cincinnati Milacron	29.0 (32.2)	50.0 (32.2)	42.5 (21.0)	42.5 (16.2)
DeVilbiss[2]	5.0 (5.5)	6.5 (4.2)	15.0 (7.4)	17.5 (6.7)
Asea Inc.	2.5 (2.8)	9.0 (5.8)	13.5 (6.7)	19.0 (7.2)
Prab Robots Inc.	5.5 (6.1)	8.2 (5.3)	8.5 (4.2)	11.0 (4.2)
Cybotech[3]	—	—	10.0 (4.9)	11.0 (4.2)
Copperweld Robotics[4]	3.0 (3.3)	3.5 (2.3)	4.8 (2.3)	4.8 (1.8)
Automatix Inc.[4]	0.4 (0.4)	3.0 (1.9)	8.5 (4.2)	20.0 (7.6)
Advanced Robotics	1.7 (1.9)	0.8 (0.5)	7.0 (3.5)	8.5 (3.2)
Nordson	0.8 (0.8)	2.5 (1.6)	5.0 (2.5)	6.5 (2.5)
Thermwood[5]	—	1.0 (0.6)	3.3 (1.6)	3.8 (1.4)
Bendix	—	—	2.8 (1.4)	6.0 (2.3)
GCA	—	—	2.0 (1.0)	7.5 (2.9)
IBM	—	—	1.5 (0.7)	8.0 (3.0)
GE[6]	—	—	1.8 (0.9)	3.0 (1.1)
Westinghouse (see note 1)	—	—	0.8 (0.4)	4.0 (1.5)
U.S. Robots	—	—	1.3 (0.6)	4.0 (1.5)
Graco	—	—	1.3 (0.6)	3.0 (1.1)
Mobot	0.8 (0.9)	0.6 (0.4)	1.7 (0.8)	2.0 (0.8)
GM/Fanuc[7]	—	—	3.0 (1.5)	8.0 (3.0)
American Robot	—	—	0.1 (0.1)	1.5 (0.6)
Textron	—	—	—	0.8 (0.3)
Nova Robotics	—	—	—	0.8 (0.3)

Source: Bache Halsey Stuart Inc., Robotics Newsletter
*1982 and 1983 figures are estimates
Notes on Table 1:
1. Subsidiary of Condec Corporation. A tentative agreement would transfer ownership to Westinghouse, but sales figures have been kept separate pending an official outcome.
2. Subsidiary of Champion Spark Plug, Inc.
3. Joint venture between Ransburg Corporation and Renault (France).
4. Robot sales estimates include sales of vision systems.
5. Includes some private label sales.
6. Excludes internal sales.
7. Includes most sales to GM (approx. 90% of total).

Table 18 (continued)
U.S. Robot Producers: Domestic Sales (mil $)
and Market Shares (%)

	1980	1981	1982*	1983*
Control Automation[8]	—	—	0.2 (0.1)	0.8 (0.3)
Machine Intelligence[9]	—	—	—	3.0 (1.1)
Intelledex[8]	—	—	—	1.5 (0.6)
Other	1.5 (1.7)	2.0 (1.3)	3.0 (1.5)	4.5 (1.7)
TOTAL	90.0 (100)	155.0 (100)	202.5 (100)	270.0 (100)

8. Excludes vision system sales.

9. Includes robots and systems; excludes stand-alone vision systems and vision portion of robot sales

The Consumers

As robot capabilities expand to include greater numbers of industrial tasks, the application of IRs grows ever broader. The most important early users were automobile producers, but improving technology and cost effectiveness have stimulated a gradual diversification of demand into a wider variety of manufacturing activities, as illustrated in Table 19.

A Summary of the U.S. Robotics Market

Robot sales of American vendors topped $150 million in 1981, more than a 60% increase over the 1980 figure. Both established

Table 19
Industrial Robot Users (partial listing)

General Motors	Northrop	Westinghouse
Chrysler	General Dynamics	General Electric
Ford	Int'l Harvester	IBM
Boeing	Caterpillar	Texas Instruments
McDonnell Douglas	Deere	Massey Ferguson
Lockheed		Briggs and Stratton
		Doehler Jarvis

Source: Robotics Newsletter, Bache Halsey Stuart Shields Inc. 4/23/80

firms and a plethora of new entrants shared in this vigorous expansion. While few expect the industry to grow at so startling a rate in the years ahead, most analysts concur that an impressive 35-40% per year figure should prove sustainable through the 1980's. This would result in 1990 sales of over $2 billion. Table 20 summarizes both current and forecasted U.S. sales and production figures.

In terms of units, output should also rise rapidly. The estimated 1981 total of 2,100 units surpassed the 1980 level by 44.8%, with future increases expected to run between 37% and 40%. By 1990, production could exceed 30,000 units.[2]

Table 20
U.S. Robotics Industry

	1980*	1981	1982	1983	1985	1990
Sales (mil$)	90	155	205	270	540	2070
Production (units)	1450	2100	3075	4000	7715	31350

Source: Robots VI Conference, 3/2/82.
*Paul Avon of Daiwa Securities America, using a slightly different robot definition, places U.S. sales at 100.5 million and U.S. production at 1269 units.

Two countervailing trends underlie both sets of figures. On the one hand, improving technology and production efficiency (as seen in the semiconductor and computer industries) should stimulate intense price competition in specific product areas, effecting an eventual "shake-out" of the weaker robot manufacturers. On the other hand, as robot sophistication grows, high-grade products will comprise an ever increasing share of industry output and sales, restoring somewhat the sales/production (or price/unit) ratio. The Japanese have made a concerted effort to quantify this trend, and their estimates should be representative of patterns observed in the U.S. and worldwide:

2. B. Sallot, Director, Robot Institute of America, and "An Overview of Artificial Intelligence and Robotics, Volume II-Robotics", National Bureau of Standards, March 1982.

	1980		1985		1990	
	% of unit prod'n	% of value	% of unit prod'n	% of value	% of unit prod'n	value
"High-grade" robots (having instruction retrieval, sensory, and reader functions)	7.4	30.5	18.5	44.2	24.8	55.2
"Low-grade" robots (simple task-repetition capabilities only)	93.6	69.5	81.5	55.8	75.2	44.8

Source: BIE

A Note on U.S. Government Policy in Robotics

U.S. government policy contains few measures aimed specifically at the robotics industry. None of these addresses the demand side of the market, so that robot use derives from traditional economic considerations. The programs in place bear only upon robot producers, and the connection is often indirect.

Two public functions should have a direct impact upon future robotics development:

A. federal standards-setting activities (still in their earliest stages);

B. government-sponsored research within universities and public organizations (National Science Foundation, NASA, etc.);

Funding in these areas totaled $15-20 million in FY 1981.[3] Other more general provisions might also serve to encourage robot manufacturers. These include:

A. liberalized tax treatment of corporate R&D expenditures (since 1981);

B. recent revisions in the time structure of the capital gains tax

3. Includes unclassified Dept. of Defense research. *Source*: Wm. Gevarter, An Overview of Artificial Intelligence and Robotics, NBS, March 1982.

and a liberalization of stock options programs (both aimed at improving the availability of venture capital);

C. the 1982 Small Business Innovation Development Act, which restructures the competition and follow-up for public research contracts.

Information is not yet available on the frequency with which these broader measures have been utilized, but they will likely have a positive influence on the industry's growth.

A Summary of the World Robotics Market

Aggregate Trends

Worldwide robotics sales passed the $1 billion mark in 1981 and could near $10 billion by the year 1990. Production (in units) will also expand at an impressive rate, with output expected to climb from 15,000 to near 135,000 robots/year over the same period. Even these conservative figures imply that the U.S. market will steadily increase in relative importance, comprising 20% of world demand by the end of the decade (vs. its present 13% share).

But growth in robotics will be shared by all the advanced industrialized countries. Most notable is the fact that, for at least the next five years, Japan will continue to produce and employ the majority of the world's output. At the same time, Western Europe—especially West Germany, the U.K., Sweden, France, and Italy—will make up yet a third market of roughly the same magnitude as the U.S.

These projected patterns for world robotics development are summarized in Table 21, "Forecasted Growth of Annual Production", and "World Market Shares by Country - 1980, 1985, and 1990". There then follows a country-by-country analysis of the main overseas robotics markets, including an assessment of public policy towards the robotics industry in each case.

Table 21
Forecasted Growth of Annual Robot Production

	1980		1985		1990	
Units	**Value**	**Units**	**Value**	**Units**	**Value**	
World 7500–8500	$660 mil	52,000–56,000	$3.4–3.5 bil	130,000–140,000	$9.5–10.0 bil
of which Japan		31,000	$2150 mil	57,500	$4450 mil
United States		7,700	$ 445 mil	31,300	$2100 mil
West Germany		5,000	$ 360 mil	12,000	$ 950 mil
United Kingdom[1]		3,000	$ 200 mil	21,500	$1420 mil
Sweden		2,300	$ 90 mil	5,000	$ 180 mil
Italy[1,3]		1,250	$ 75 mil	3,500	$ 225 mil
Norway		1,000	$ 50 mil	2,000	$ 103 mil
France[1,2]		1,000	$ 50 mil	—	$ 150 mil

Sources: Robot Institute of America, plus those cited in Table 21 footnotes.

[1] Values based on European average of $65,000/unit.

[2] Projection using L'Usine Nouvelle estimates for 1981 market and future production/growth.

[3] Projection using Le Progres Scientifique data from 1979 market estimates of future production/growth.

119

Figure 3
World Market Shares
by Country—1980,
1985, & 1990

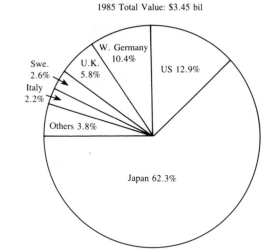

1985 Total Value: $3.45 bil

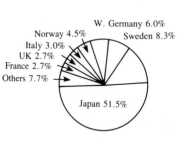

1980 Total Value: $660 mil

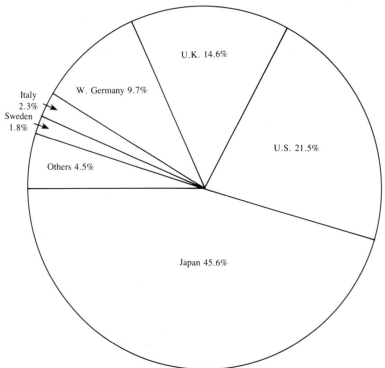

1990 Total Value: $9.75 bil

Sources: RIA, L'Usine Nouvelle, BRA, Veckans Affarer, Capital. Certain Totals represent average values across multiple forecasts & IR definitions.

ROBOTICS

Country-by-Country Analysis

Japan

Although the Japanese did not themselves develop the fundamental technologies involved in the industry, they did launch the robotics revolution by first pursuing the widespread use of industrial robotics in a factory setting. Four basic features of Japan's economic environment contributed to its unique willingness to explore robot-based manufacturing techniques: an unusual concern for productivity, a heavy emphasis upon product quality, a cooperative relationship between labor and management, and an organized program of government incentives and support.

As in the U.S., the automakers proved the catalysts, generating the first reliable, private demand for industrial robots. But again, functions and applications broadened quickly, reflected in Table 22, a list of principal Japanese users of IRs:

With this growth has come similarly rapid diversification and specialization among robot-makers in Japan. Table 23 brings together the leading Japanese suppliers and indicates which IR applications have become their production emphasis:

The domestic market formed by these Japanese consumers and producers has swelled dramatically over the last 5 years, and the surge should continue at current rates through 1985. Although growth may slip somewhat thereafter, Japan will have established standards for annual output and production value that the U.S. will not attain until 1990. While this five-year "lead" need not imply Japanese technological superiority in the robotics field (indeed, the U.S. could well remain the center for innovation), it does illustrate the relative maturity of a Japanese industry well-sprinkled with high-tech conglomerates. Until recently, the emphasis on robotics in Japan had taken a fairly simple form—to put in place, as quickly as possible, as many IRs as industry could be induced to absorb. The logic behind this strategy: 1) to cultivate a steady reliable demand for robots by introducing a wide variety of manufacturers to their use, and 2) to stimulate the rise of a full-fledged robotics industry, with the stable capital/financial base and production capabilities necessary to meet future growth at home and abroad. By and

Table 22
Shipments to Industry
(% distribution by value)

By Sector	1980	By Function	1980	1990 (proj.)
Auto Manufacturing	36	Machining	22.8	13.1
Electronic Equipment	21	Spot Welding	11.4	7.5
Synthetic Region Processing	10	Assembly	9.8	21.7
Metal Product Manufacturing	5	Arc Welding	7.3	10.3
Metal Processing Machinery	4	Plastic Molding	7.2	3.3
Non-Ferrous Metals Manufacture	3	Control	4.9	10.0
Steel	1	Painting	4.9	4.9
Export	3	Metal Cutting/Stamping	4.9	4.5
Others	8	Pressure Casting	3.3	1.2
	100%	Casting	1.6	2.4
		Heat Treatment	1.6	2.4
		Forging	1.6	1.3
		Other	17.3	17.6
			100%	100%

Source: Sangyo Yo Robot Guidebook, 10/81; Japanese Industrial Robot Asssociation.

large, the Japanese have succeeded in these respects, and they have done so well in advance of their overseas competitors.

Of course, the inevitable result of this approach was an early, lopsided strength in the low-technology end of robot manufacturing and application. But now that this groundwork has successfully been laid, the Japanese have set their sights in new, though predictable, directions:

1. Technological leadership in the robotics field—The "Japanese Government in Robotics Development" section below details the major government/industry programs aimed at advancing IR research and exploring new IR applications.

2. A share in the growth of overseas markets—Because of its disproportionately large domestic market, Japan has to date exported less than 3% of its robot production. But as the rest of the world follows in its footsteps, Japan hopes to serve as a major international source, the presently stated goal being to export 15% of to-

Table 23

The Top Japanese Robot-Makers in the Major Market Segments

Field		Ranking
Spot Welding	1)	Kawasaki Heavy Industries (>50% share)
	2)	Mitsubishi Heavy Industries
	3)	Nachi Fujikoshi
Arc Welding	1)	Yasukawa Elec. Manuf. (appr 50% share)
	2)	Osaka Transformer
	3)	Shinmeiwa Industry
Spray Painting	1)	Kobe Steel (appr 50% share)
	2)	Mitsubishi Heavy-Industries
	3)	Tokico
Assembly	1)	Fujitsu Fanuc
	2)	Sankyo Seiko
	3)	Nitto Seiko
Press	1)	Aida Engineering

Source: Bache Halsey Stuart Shields, Inc.

123

tal output by the year 1985. (Further discussion of these points also in the section on "World Robotics Trade".)

Their prospects for success in attaining these two new objectives are difficult to judge. Technological superiority may prove particularly elusive, although a shift to greater production of advanced robots appears certain (See Footnote #2). On the other hand, the Japanese should at no point find themselves eclipsed by innovations among their U.S. and European competitiors. Their research programs are intensive and well-organized, based upon a solid background of experience in the robotics field.

Exporting 15% of production represents· an ambitious target for 1985 (it would require 20% control of all overseas markets by that time), but growing Japanese participation in international licensing agreements and joint ventures could bring these penetration levels within reach. Most observers then expect Japan to peak in terms of the relative international position of its robot producers. The key reason for such a forecast:

1. As the domestic market matures, growth rates should settle into the 20-30% per annum range (value of sales in yen);

2. foreign competition will have come into its own, providing a very stiff challenge for surging demand in the West;

3. technological advances should gradually render Japan's early lead less and less significant; and

4. within a few years, the difficulties for the Japanese in establishing the necessary international marketing, service, and customer support networks may constrain further improvements in their competitive position. ·

The Japanese Government in Robotics Development

In a manner typical of its past targeting programs, the Japanese government has assembled an industrial policy package aimed at robotics development. Its key provisions are designed to coordinate and advance the research efforts of current robot manufacturers while broadening and accelerating IR application throughout the manufacturing sector. The major programs currently in place include:

1. A seven-year, Y30 billion ($150 mil) research project on the development of intelligent robots, sponsored and funded by the

Ministry of International Trade and Industry (MITI). It will extend through early 1989.

2. Formation of the Japan Robot Leasing Company (JAROL), which, with funding via the Japan Development Bank, offers subsidized leasing arrangements to a wide range of potential customers. Contracts extended in its first year (FY 1980) of operation reached $5.8 million; further liberalization of terms in 1981 should cause this number to grow rapidly.

3. Extra depreciation allowances on robots (an additional 12.5% per annum for three years), enabling firms to depreciate 52.5% of the purchase price in the first year alone.

4. Direct, low-interest loans through the Small Business Finance Corporation to medium and small-scale manufacturers purchasing robots. The FY 1980 budget included Y5.8 billion ($29 mil) for this purpose.

5. MITI support (value unknown) for development of a fully automated factory for small batch engineering components and assembly.

6. 20% MITI funding (Y2.6 billion or $13 million over FYs 1977-83) for development of a Flexible Manufacturing Complex provided with lasers.

The exact contribution of these measures to the progress of Japan's robotics industry will always prove difficult to assess, but there can be no doubt that such significant incentives have had some positive effects. While the broader economic factors discussed earlier also deserve considerable credit, Japan's sudden robotization clearly drew much of its energy from both the tone and substance of government assistance.

W. Germany

West Germany ranks a solid third behind the U.S. and Japan in robot usage. Applications parallel closely the patterns observed elsewhere, with the automotive industry heading a list of diverse, but at present less significant, customers. Robot manufacturers also include a typical assortment of machine tool, electrical equipment, and transportation companies:

Volkswagen	Keller & Knappich (KUKA)	Koenig
BMW	Pfaff Industrie Machinen	Feiss
Bosch	Roth Electric	
Siemens AG	VFW Fokker	

Though only 1400 robots were in place in Germany at the end of 1981 (vs. 4150 in the U.S. and 14,000 in Japan), sales and output are expected to grow at nearly 50% per year through 1985, before slowing to more modest 20% rates near the end of the decade. These projections and data indicate that Germany will indeed emerge as a major robotics market by 1990.

Government policy, while supportive of robotics development, has remained less comprehensive in scope than its Japanese counterpart. Apparently, no demand side measures are in place; instead, funding has been channelled either directly into research (both basic and applied) or towards the training of technical/engineering personnel. In any case, analysts doubt that public programs will decisively influence the German robotics industry. Rather, its greatest strength should stem from the considerable manufacturing capability of the market's recent entrants (see the above list). The presence of these major firms could eventually place the Germans in a prominent position across much of Western Europe.

Sweden

Over the last few years, Swedish manufacturers have established a solid reputation for production of highly sophisticated industrial robots. While this specialization in advanced technology products has resulted in a lower volume of shipments than among other major robot-producing countries, it has also enabled leading Swedish firms to carve out a reasonably secure niche in an increasingly competitive world market. Both domestic output and consumption should increase according to broader world trends, again with growth easing somewhat in the late 1980s. Projections are summarized in Table 24.

Because of their relative specialization, certain Swedish firms

Table 24
Sweden's Robot Production*

	1981	1985	1990
Value	$50 mil	$90 mil	$190 mil
Units	800	2300	5000

Sources: International Survey Institute; Robot Institute of America.
*Output includes a current 35% share of the Western European market.

should perform well on an international level. This is particularly true of the dominant manufacturer, ASEA (which in 1981 accounted for over 500 units, 60% of Sweden's IR production at a per unit value of K400,000, and of its nearest competitor, Electrolux (120 units at K250,000/unit). As in West Germany, most government assistance flows to manufacturers in the form of R&D funding, with some additional support through public technical institutes and broader export-oriented programs. Though details of these practices are scarce, it again seems unlikely that they will play critical roles in the continued development of Sweden's robotics industry. More significant for this export-oriented business will be the outcome of the international agreements negotiated by the companies involved. (See the International Agreements section below for discussion.)

The United Kingdom
Hampered by serious union problems, Britain had until recently moved slowly in the robotics area. But now the Thatcher government appears to have given considerable priority to robotics development, and consequently a solid increase in both robot application and production is expected over the next several years. In fact, because of its slow start in the field, and because of the broader economic advantages of automation under conditions of labor management confrontation, the U.K. could well sustain a 35%–50% growth rate through 1990. This kind of success would leave Britain behind only the U.S. and Japan in robot population by decade's end.

Table 25
The British Robotics Market

Producers in the UK	% of 1981 IR Population	Applications (no. in use)
Unimation Europe	Domestics 26%	Welding (100)
Hall Automation (GEC)	Imports 74%	Painting/Finishing (70)
Lamberton Co.	of which:	Machine Loading (150)
Sceptre Eng. Design	(Japan . . . . 8%)	Assembly and Misc Uses (50–80)
Modular Technology	(the U.S. . . . . 30%)	
Pendar Group	(Europe . . . . 36%)	Total population (mid-1981): 370–400

Source: British Robot Association

To date, subsidiaries of U.S. firms have been responsible for much of the UK's robot production, but new entrants should quickly diversify the competition, and together they should steadily reduce import penetration. Robot usage, on the other hand, has evolved along more traditional lines.

But the most striking feature of the robotics situation in Britain is the ambitious network of government support for producers and consumers alike. The following list includes only the most prominent programs:

1. The Science Research Council's £2.5 million ($5 mil) project studying future generations of robots, including such capabilities as automated micro-assembly, free moving trucks, and software simulation.

2. CAD/CAM and CAD/MAT schemes include "awareness and demonstration" funding within their £27 million ($54 mil) budget.

3. The Product and Process Development Schemes (PPDS) provide:

    a. free advisory visits for prospective robot users;

    b. grants for up to 50% of cost for robot feasibility studies;

    c. grants for up to 25% of development and installation;

    d. grants for up to 25% of costs of projects "involving the design and development by UK manufacturers of a new IR and associated equipment."

4. National Engineering Laboratory and selected trade associations have received roughly £650,000 ($1.3 mil) per annum for robot-related studies.

5. Subsidized financial backing is also available from regional development offices for robot production ventures.

Such an extensive program of incentives could certainly have a significant impact upon the pace of robotics development in the UK. The unions' interest in and success at forestalling robot usage remains difficult to assess. Perhaps the more critical question asks the fate of these targeting practices should a change in government occur. But for the present, at least, Britain's public commitment to robotics appears unquestionable and, consequently, its near-term market growth seems assured.

France

France also finds itself in the early stages of robotics development. Production falls either to the major Renault subsidiary, ACMA, or to a smattering of small, more specialized firms. Usage remains largely confined to automobile manufacturers, although increased application is expected from other industries in the near future.

French Robot Producers

ACMA (Renault)	Languepin
AOIP	Scialcy
Afma-Robots	La Calhene (CEA)
Climax (Compair-UK)	

The French government is clearly committed to promoting the nation's robotics industry, and many of its policies are aimed at supporting R&D programs on the production side of the field. These range from basic research on new robot types to more comprehensive studies of robot manufacturing procedures, and their sponsorship derives from several sources:

1. The Committee for Development of Strategic Industries (CODIS);

2. Institut National de Recherche d'Informatique et d'Automatique (INRIA);

3. The Advanced Automatization and Robotics Group, with participation from the National Scientific Research Center (CNRS) and the General Delegation for Scientific and Technical Research (DGRST);

4. The National Robotics Plan involving industrial and research leaders; and

5. The Robotics Commission, affiliated with Mechanics Department of the French Ministry for Research and Technology.

But an effort has also been made to stimulate demand for industrial robots, both by intensifying their use in presently automated areas and by encouraging potential customers in experimenting with new applications. Some Fr160 million ($29 mil) were included in the 1981 budget for this latter purpose, while public R&D allocations totalled some 91 million francs ($17 mil).

ROBOTICS

The long-term prospects for France in the robotics field are particularly difficult to evaluate. Though modest growth is expected, demand outside the transportation sector has yet to take shape. Most of the smaller robot manufacturers find themselves in a fragile business position, facing the challenge of both ACMA (Renault) and heavy import penetration (up to 50%). Finally, government programs provide only tenuous support for many, in part due to understandable public ambivalence towards promoting a labor-replacing product, in part due to the capriciousness of economic policy in times of dramatic political change. Apart from ACMA and its automotive customers, French robotics remains decidedly in the formative stages. Of course, a quick history of any nation's market underscores the fact that this means only a 2-3 year lag, and recent growth in France has been impressive. But at present, few analysts expect the French robotics industry to be a major force on the world scene in the foreseeable future.

Italy

The Italians started early and started strongly in the robotics field. Four major producers share most of the nation's output, and, while domestic consumption remains modest, an impressive export per-

Table 26
Italy's IR Markets*

	1975		1979		Annual Growth Rate	
	Units	**Value****	**Units**	**Value**	**Units**	**Value**
Production	60	0.7	420	20.0	62.6%	131.2%
Imports	40	2.8	40	2.3	0.0%	(5.0%)
Exports	20	0.2	160	7.8	68.1%	151.6%

Source: Le Progres Scientifique.
*Data includes manual manipulators (approximately 50% of 1979 production represents robots narrowly defined).
**Value given in billions of current lira.

formance has assured steady growth in total demand. Recent data is summarized below.

The enviable record of the late 1970s will, however, prove difficult to sustain through the 80s. First, volume is still low, and the capital requirements of continued expansion could outstrip the sector's financial capabilities. Second, as overseas producers gather momentum, penetration of the foreign markets upon which Italy now depends should become increasingly difficult. But despite the fact that the Italian robotics industry has arisen with a minimum of public assistance (in fact, a lack of research funds may constrain further growth), it promises to challenge its American, Japanese, and European competitors for some time.

Others

In *Norway*, the key industry figure is Trallfa, producer of approximately 80% of the world's stock of painting robots. This type of specialization (Trallfa's impact is negligible in other areas) may establish a pattern for smaller firms in selected areas of the robotics industry.

In the *Eastern bloc*, serious technological disadvantages have left production some 10 years behind in terms of robot sophistication and capability. the USSR, Poland, and Bulgaria have sought to accelerate factory automation and have deployed a fair number of robots in industrial settings. But these countries should provide neither challenge nor opportunity in the near future.

World Trade in Robotics

Although the traded fraction of the world's robot production is significant (10-20%), both the volume and value of international flows remains comparatively small (roughly $100-$200 million). Unfortunately, statistical information in these areas is woefully incomplete, and the data, where available, is rarely comparable across a useful range of countries. The development of a universal classification system that explicitly accounts for trade movements in robotics will surely require several years. In the meanwhile, analysis will have to rely on occasional estimates and accept an added measure of uncertainty wherever the numbers prove inadequate.

The question of the U.S. trade balance in robotics poses perhaps the greatest challenge of all. The Bureau of Industrial Economics is currently working with the Census Bureau and industry representatives to develop SIC categories reflective of robotics as a discrete sector. Meanwhile, the TSUS (Tariff Schedule of the United States) code makes two specific provisions for IR imports, with various other basket categories incorporating the remainder of robot arrivals. And Schedule B, the primary export classification system, provides no help whatsoever on that side of the ledger: export trade information is generally included with the data for machine tool shipments, with functional breakdowns providing minimal guidance. The relevant categories and net results are summarized in Table 27.

Analysts generally agree that the U.S. enjoys rough balance to

Table 27
US Robotics Trade

Imports (TSUS)	1981	1982 (est)
Materials Handling Robots (664.1005)	$2.8 mil	$ 4.5 mil
Welding Robots (683.9005)	$4.1 mil	$ 8 mil
Robots in "Basket" Category (678.50)*	$6 mil	$10 mil

Note: None of these classifications includes metal-working, spraying, cutting/drilling robots or accessories or parts.

Exports (Schedule B)	1981	1982
Robots in "Lifting, Handling, Loading & Unloading Machines, N.E.C.", (664.1078)	Specific Estimates within these Categories not	
Robots in "Welding & Cutting Machines N.E.C." (683.9040)	Available. Classification Headings	
Robots in "Other Machines" (678.5090)	Provided for Reference Purposes.	

Note: None of these classifications includes metal-working, spraying, drilling robots or accessories or parts.

*All estimates from US Customs Service via Int'l Trade Commission.

Table 28

Firms	Home Country	Product Emphasis
ASEA	Sweden	Arc & Spot Weld'g, Mat'l Handl'g
Trallfa/Devilbiss	Norway	Painting/Spraying
Olivetti	Italy	Assembly & Others
Kuka	West Germany	Welding
Fujitsu Fanuc		
Hitachi		
Kawasaki	Japan	Across most IR applications
Mitsubishi		
Yasukawa		

perhaps a slight surplus in robotics trade, with current import penetration levels at about 10-15% of consumption and overseas shipments near 15-20% of output. The foreign firms accounting for the bulk of US imports include:

Meanwhile, US exports head almost exclusively to Europe through Unimation (22-25% market share in EC) and Cincinnati Milacron (6-8% EC market share).[4] Their main destinations include Great Britain, France, and West Germany.

The foreign sector plays a more prominent role (in percentage terms) among the European countries, where Sweden, Norway, and Italy also enjoy a robotics surplus, while France and the U.K. (along with other smaller nations) remain net importers. Of course the Japanese, despite exporting less than 5% of production, also posted an impressive surplus in terms of both units and value traded.

Trade in robotics should grow more rapidly than the field itself. The sudden proliferation of marketing agreements should dramatically increase international shipments of IRs and their parts. Since much of the scramble is aimed at the American market, a surge in U.S. robotics imports near-term can be expected. The Japanese have put forward objectives whose attainment will require a 80%-100% per annum rise in exported value (even more in terms of units) over

4. Market shares include European production of US subsidiaries.

Table 29
Japanese Trade in Robotics

Exports (1981)	Imports (1981)	Balance
1,170 units	2 units	+ $15.0 mil

Source: Financial Weekly (UK), 1/4/82, p.2; Canadian Data Systems, 4/82.

the next 5 years. Corporate specialization in particular robot types will also contribute significantly to trade flows. Countervailing forces include the importance of customer service and support (a modest advantage for locally-based manufacturers), the interference posed by existing targeting practices, and the possibility of more prohibitive trade restrictions where serious imbalances develop. But for at least the next few years, the factors favoring an increase in trade should outweigh the disincentives, resulting in a rapid increase in international shipments of IRs and associated products. Most of this intensified international competition will center around the struggle for market share in the producing countries. But important tests will also be faced in some of the newly emerging third markets. South Africa, Korea, Taiwan, Singapore, and Brazil are only a few of the nations likely to explore IR usage but unlikely to produce robots in the near future.

Key Issues Affecting
U.S. Competitiveness in Robotics

International Agreements
Between Private Firms

A trend for which there already exists clear evidence, and which will decisively affect the industry's future, is the growing preponderance of international agreements. These run the gamut, from marketing accords to licensing compacts to full-fledged joint ventures.

The majority represent pairings where a) the foreign firm seeks quick access to a particular national market with immediate sales and service support, and b) the host firm hopes to absorb the technology embodied in its partner's product while making a name for itself among robot purchasers. Agreements of this type are found below in Group A.

Other cases are much more balanced in terms of their technology exchanges and the resulting production/sales distribution. These two-way accords are cited in Group B.

Finally, a third type of collaboration couples a major robot producer with a major robot consumer, streamlining the flow of IRs between two nations and two firms. The prime example to date is listed under Group C. A fairly recent phenomenon, this may start a trend in multinational accords.

The straightforward effect of these agreements should be to:

1. internationalize the robotics industry, effectively joining world's major markets (Japan, the U.S., and Europe);
2. accelerate the diffusion of robot technology; and
3. facilitate both new entry and trade.

Japanese and European firms now have unprecedented access to the U.S. market. Selected American robot manufacturers also enjoy new opportunities in Europe, though Japan will remain quite difficult to penetrate. Several major U.S. companies can now draw on others' production capabilities in their attempts to establish a position of their own. A dramatic realignment of the industry has now been accomplished, but its full consequences, especially for the eventual competitiveness of the nations represented, will only gradually be recognized and understood.

Standards and Regulation

Responsibility in the U.S. for this area is shared by the National Bureau of Standards (NBS) and the Occupational Safety and Health Administration (OSHA). The former organization has initiated a standards-setting program to develop criteria in two areas:

1. The interface between robot and control mechanism, and
2. the mechanical tolerance features of robot performance.

Table 30
International Agreements Between Firms

Group A—Between Producer/technologist and a marketer/servicer	*Group B*—More balanced exchanges of technology and production/followup activity	*Group C*—Producer-Consumer Agreements
Komatsu Ltd.→Westinghouse	Yasukawa↔Machine Intelligence Corp (MIC)	Fujitsu Fanuc→General Motors
Mitsubishi Electric→Westinghouse	ACMA-Renault (France)↔Ransburg	Note: Similar contracts of this sort occur involving national subsidiaries of foreign firms, instead of formal international pacts with overseas producers.
Sankyo Seiki→IBM	Siemens AG(FRG)↔Fujitsu Fanuc to form General Numeric (US)	
Hitachi→General Electric[2]		
Hitachi→Automatix		
Nachi Fujikoshi→Advanced Robotics		
Volkswagen AG (Germany)[3]→GE[2]		
DEA (Italy)→General Electric		
Olivetti (Italy)→Westinghouse		
Trallfa (Norway)→DeVilbiss		
Trallfa (Norway)→Kobe Steel		
Unimation→Kawasaki[2]		
Unimation→Nokia (Finland)		
Prab Robots→Murata Machinery		
Prab→Fabrique Nationale (Belg.)		

Notes: [1]Sale for resale accord.
[2]Definitely involves some production under license.
[3]Non-Japanese partners marked by country of origin.

NBS officials expect that a voluntary code will take effect in the late 1980s, with amendments to follow as robot technology continues to evolve.

OSHA standards are more reactive in nature, compiled in response to industrial accidents and documented malfunctioning hazards. Because rules are generally structured according to either workpost or mechanical component, guidelines applicable to robots as a separate group or self-contained category have yet to materialize. Thus far, the safety record of robots has been outstanding; as a result, they have not been involved in any significant OSHA cases. Future regulations should above all reflect the reliability of robots' on-the-job performance.

Thus far, the U.S. government has paid little attention to standards and regulation in the robotics field; but even its relatively minor efforts represent a more concerted approach than one finds overseas—with one important exception. The Japanese have adopted a series of product standards designed to coordinate research efforts across the industry, improve the compatibility and interchangeability of Japanese-made IR components, and facilitate the integration of robots of differing functions and sources into larger and more flexible manufacturing systems. Though perhaps at some cost in terms of innovation, this combined policy of specialization and "rationalization" has channelled both product and applications research efforts into a more cumulative robot development process.

Labor and Automation

Increasing attention has been focused upon the implications for the labor force of recent automation trends. The advent of industrial robots, because they so directly replace manpower, has made these concerns more pressing than ever. A recent Carnegie-Mellon Robotics Institute study found that present generation robots could replace approximately one million manufacturing workers, with an additional 7-9 million within reach of the next group of high-grade machines. The Rand Corporation has forecast that as little as 2% of the workforce will be employed in manufacturing in the year 2000 (vs about 20% today), and German research asserts that, unlike computers, robots will eliminate five jobs for every one they create.[5]

But despite these ominous forecasts, most union leaders have decided to "work with" the automation movement, recognizing that to reject it would mean a serious lag in industrial productivity and, eventually, the loss of even more jobs. Yet the problem of worker attitude is only one, though perhaps the most important, of the impediments to widespread robot use. The next problem, reabsorption into the labor force, could require substantial retraining and relocation. While sizeable shortfalls are expected in the supply of high-technology skills (computer technicians, for example), the dramatic transition required to fill such slots may prove difficult to effect. Programs designed to facilitate these shifts will have to expand and mature rapidly to deal with the greater numbers and stiffer challenges involved.

Obviously, the benefits of automation are many, including the assumption of most hazardous jobs and tedious/repetitive tasks, improvements in output quality and uniformity, and uninterrupted, around-the-clock production capabilities. Rising wage levels and declining robot prices have combined with steadily improving product performance to provide inexorable economic justification for robot purchases. But widespread robot use will require fundamental sociological adjustments on several levels—economic, educational, psychological, and cultural. Nevertheless, the powerful emerging trends documented in this profile indicate that these adjustments will have to be made.

Keys to Future Success in Robotics

Long-range success in the robotics industry would seem to depend on several factors:

First: In terms of the product itself, five basic forces appear to drive the marketplace and describe the robots of the future: 1) lower cost, 2) smaller size/greater strength, 3)improved sensory (aural,

5. *Source:* The Industrial Robot. 3/82, P. Kalmbach et al, "Robots Effect on Production, Work, and Employment."

tactile, and visual) technologies, 4) broader capabilities through "Systems Integration", and 5) a background of effective marketing strategies in the face of crowded competition.

Second: Robotics firms will have to cultivate a diverse set of strengths: 1) customer support, 2) strong R&D programs, 3) software enhancements, 4) understanding of the factory environment, 5) marketing strength, and 6) financial strength.[6] As the field races towards Flexible Manufacturing Systems (FMS) and greater levels of robot intelligence, producers will have to combine fertile imaginations with the more traditional attributes of sound corporate standing, absorbing all advances in source industries (semiconductors, software, etc) while anticipating and meeting future demand directions. Even a brief lapse in one of the above areas could prove fatal under the competitive pressure expected to prevail.

A Final Note on U.S. Prospects

The United States robotics industry probably remains the world's technological "leader", but this delicate margin will little comfort those who appreciate its limited significance for long-term international competitiveness. First, the field changes too rapidly for any such edge to have permanent significance. Second, information diffuses too easily for an advantage of this type to endure, even without further innovation. And finally, competitiveness in high-growth sectors is a function of too many other industry characteristics to rest on such scanty laurels. The Japanese enjoy undisputed superiority in terms of their producers' experience, capacity, financial strength, and market position. Only the recent arrival of major American firms (GE, Westinghouse, IBM, etc.) has even opened the prospect of a general American challenge in these critical areas of distinct Japanese advantage. Of course, should competition become a battle between industrial policies, the resulting inefficiencies could skew market development in ways impossible to predict.

6. For further discussion of product and company features. see "Robotics Newsletter", No.1, April 1980, Bache Halsey Stuart Shields, Inc.

But at the same time, several factors will brighten the future of the U.S. industry:

1. Robotics has gradually earned its status as a legitimate industry (reinforced most recently by the interest of the giant firms mentioned above). This, combined with the rising certainty of rapid growth, should now ensure a continuing flow of the venture capital necessary for further progress and expansion.

2. R&D funding, from both private and public sources, should increase at rapid rates over the next few years, especially as more and more companies search for their technological niche.

3. The larger, more integrated firms now arriving on the scene should succeed in applying their technological expertise in related areas to robotics development. The general strength of the U.S. in software could prove a particularly significant edge in the years ahead.

4. Rapid robotics growth in all of the advanced industrialized countries should provide ample entry opportunity for local firms while limiting a particular nation's ability to move strongly out of its domestic market. In the near-term, this trend should favor the US.

5. Insofar as marketing becomes a critical factor in the U.S. market, American firms should enjoy somewhat of an edge at home, though this same condition could limit US success overseas.

6. Greater government awareness of the importance of high-tech development for the future health of the US economy should make for greater federal sensitivity to the public policy concerns of the high-technology sector.

The United States' international standing in robotics should gradually improve, but for the present, the industry places a fairly distant second. Uncertainty permeates most forecasts, yet one prediction is certain: Japanese and European competition will provide a substantial challenge to American manufacturers both in the U.S. and world markets.

Options

A Discussion of the pros and cons of proposals for USG action as recommended by a variety of sources

Domestic Demand

A first area of concern for many observers of the U.S. robotics industry is the apparent sluggishness of domestic demand for IRs. Many feel that this could have long-term implications for the international competitiveness of U.S. robot manufacturers. This reluctance on the part of potential customers is traced to several sources: the general economic recession and unavailability of capital; a resistance among mid-level administrators in plant management positions to the types of changes involved in the robotization of production processes; the fear among labor groups and the general public that the introduction of robots will eliminate jobs; and a lagging sensitivity to the problems of product quality addressed by automation.

Tax incentives specifically aimed at promoting automation (accelerated depreciation, tax credits, etc.) would directly increase robot usage across a modest range of industries, ultimately generating improvements in productivity, output quality, and industry competitiveness. However, it might introduce distortions into the existing capital/labor balance among user industries, distortions that could even in the long run prove economically unjustified.

It could divert investment capital away from other types of manufacturing activities and channel demand away from other investment-good markets.

It would establish a difficult precedent that could then serve as a pressure point for similar action in other automation-related industries (computers, telecommunications, etc.).

More extensive USG *encouragement of research and development* activity relevant for both user and producer firms (through tax policy, publicly supported projects, etc.) would address fundamental concerns for America's declining position in terms of its technological capabilities. It could rekindle more aggressive pursuit and application of new concepts in the area of industrial production including robot development and usage.

Some incentives aimed in this direction have already been introduced, most notably the credits included in the 1981 Economic Recovery Tax Act. Of course, the long-term effects and adequacy of those measures have yet to be determined, and it may be prudent to postpone further action in this area until some assessment can be made.

The accelerated depreciation schedules for R&D equipment, now in place under ERTA, should provide significant and secure incentives for corporate investment in this area. The accompanying R&D tax credits, however, may need elaboration to ensure their effectiveness.[7] Insofar as firms' decisions on R&D allocations require long-term planning and more extended lead-times, two years (the effective period of these current measures) may prove inadequate for generating a broad positive response. A longer-term provision of this type may prove desirable. A second possible shortcoming of the stepwise R&D credit may be its lack of stimulus for the young, fast-growing companies that so heavily populate research-intensive sectors, and from which an impressive proportion of technological innovation has emanated. The simple incremental approach embodied in existing legislation may provide the least benefit and incentive to many of those most active in the area of policy concern. Revisions that structure into the formula credits for a baseline, dollar amount R&D increase (on top of which the 25% schedule would take effect) might somewhat alleviate this problem.

Another set of policy developments in the R&D field involves more open interpretation of anti-trust regulations. Recognizing that some legitimate economies of scale can be realized through limited inter-firm collaboration in areas of basic research, the Justice Department has given qualified approval to the establishment of the joint venture Microelectronic and Computer Technology Corporation (MCC). While this case specifically involves only the computer industry, many observers feel that it could represent an important first step towards similar cooperative activity in other high-technology sectors. But it appears that before any field can reap the full rewards of this new understanding, the ground rules will need to be clarified and secured. The lack of detailed and defensible preconditions will likely deter many valid participants from joining a collective undertaking of this type. At this point, considerable discretion-

7. Included are both the "25% of increase" credit and the reallocation of international credits assigned under Code Section 861.

ary/interpretive power remains with the Justice Department, the courts' position on such ventures has yet to be clarified, and no protection from civil suits has been provided. In the face of such impediments, legislative action may emerge as the only mechanism able to catalyze full use of this collaborative opportunity.

Retraining programs for displaced labor would ease the dislocation fears that accompany robot utilization and help to stabilize labor-management relations in a potentially tendentious area. It would begin the process of retraining the U.S. labor force from low-tech to high-tech job skills.

Most of the problems here are practical. Retraining, to be most effective, would need to precede or at least coincide with robot deployment, and may often need to prepare workers for positions in completely unrelated sectors or geographic areas.

Incentives to companies to donate equipment to educational institutions would broaden mass exposure to robots, robot functions, and robot capabilities, reducing resistance to robot usage and transforming vague fears of automation into a specific interest in robot applications and potential. (See the Computer Equipment Contribution Act of 1983, now in the Congress.)

However, only if such a program included explicit preconditions—donation requirements ranging from instruction in design usage to simple maintenance assistance to curriculum specification—would it offer any real prospect of effectively improving public acceptance and understanding of robots and their capabilities.

Countering Foreign Competition

A second concern is the proliferation of foreign government programs aimed at robotics development within the respective countries, frequently with an eye towards an eventual export capability. These may significantly affect the international competitiveness of the U.S. robotics industry.

Comparable targeting practices could be formulated to spur development in any of several areas of the robotics industry. Certain possibilities have already been discussed in considering various demand stimuli. Others could encourage risk-taking among producers or, in the early high-growth stages of the industry, avoid undesirable waste and duplication in certain R&D areas. Presumably, the adoption of such measures could place U.S. manufacturers on an "equal footing" with their foreign counterparts.

Certain minor types of support (especially basic research in government and university laboratories) and special tax credits for R&D are already in place. More sweeping measures would require a fundamental change in the current philosophy of business-government relations in the U.S. Such revisions could also either a) shift competition in robotics from production programs to support programs as other countries in turn attempt to provide the most generous terms for development or b) precipitate the introduction of less palatable trade barriers, such as tariffs, quotas, etc. And again, industry-specific USG policies would invoke demands for equal treatment across a whole range of American sectors that feel similarly victimized.

Protecting the U.S. Market could result in an eventual dismantling of selected foreign industrial policy programs if this is accepted as the price for regaining access to the U.S. market. But, under GATT provisions, the U.S. is ineligible for protection via the "infant industry" clause. Given the vagueness of many foreign provisions and the inherent competitiveness of the industry, injury to U.S. producers may be impossible to prove. And since most imports will probably now enter under joint venture agreements involving American firms, such action would even have ambiguous short-term consequences for U.S. interests. This solution also fails to address the question of third markets—and protective diversion of foreign exports will erode U.S. market shares abroad. Finally, such unilateral action raises the specter of a full trade war, an eventuality that could seriously damage the U.S. economy across a much broader range of products and industries.

A vigorous U.S. program to *counter targeting programs* through strict enforcement of U.S. trade laws (under Section 301) could produce case-by-case agreements as to how an equitable trading environment could be restored. Historically, this has been a successful process, with only rare invocation of Executive Authority to impose

145

unilaterally reciprocal restrictions. Above all, active enforcement would lend integrity to the legal structure now in place.

Most U.S. trade laws only emphasize temporary relief and adjustment assistance where damage is found, offering little to actually discourage targeting practices. The resources required of firms to pursue trade action cases, and the often lengthy period between violation and judgment, may discourage many (particularly smaller companies) from invoking what provisions are available. While several complaints are still in decision at this time, Section 301 (the mechanism relevant for most targeting problems) would also appear not to deal with the question of third markets, and many nations may well decide that the benefits of promotion policies still outweigh the costs of American enforcement. Finally, there remains some question as to the GATT-legality of certain responses of this type.

Negotiation through bilateral channels for country-by-country *removal of the most restrictive targeting practices* could give rise to a consistent and principled U.S. strategy for dealing, in the context of individual bilateral relationships, with particularly onerous forms of support. The U.S. approach to discussions of this kind would have to take into full consideration the range of other economic, political, and security interests at stake. Because these may vary greatly from case to case, it could prove difficult to develop any set of policy positions on targeting that appears coherent and non-arbitrary.

Government Regulation

The third concern of many who observe the U.S. position in high-technology sectors is the role of government regulation in streamlining or hampering robotics development. Two types of policies receive particular attention because of their implications for the robotics industry: standards setting programs which would focus primarily on product codes in the areas of interface and safety, and export controls promulgated to limit technology transfer to the East-

146

ern bloc (such as pending COCOM restrictions on the flow of robots featuring more than three degrees of freedom).

USG policies in the area of export control will clearly be attempts to balance sometimes conflicting objectives, and any attempt to dictate a particular solution in this context would prove highly problematic. However, the upcoming renewal of the Export Administration Act does provide an opportunity for debate over the methods and content of current policy practices. As part of the review, it could be appropriate:

A. to re-emphasize the priority of technology transfers over product transfers as a guiding principle for security concerns;

B. to underscore the broad damage to U.S. commercial interests that results from perpetuating the United States' reputation as an "unreliable supplier";

C. to highlight the fact that export markets in high-technology fields also contribute to national security by expanding and strengthening the military/industrial base;

D. to urge all policy-makers involved to give full attention to the competitive interests of the relevant manufacturers, recognizing in particular the disproportionate burden that export restrictions can place on the smaller firms that have typified the industry; and

E. to formalize this advocacy role by involving Commerce Department industry and trade policy specialists in future discussions of robotics regulation.

Robotics Industry International Overview

BY JOSEPH ENGELBERGER
PRESIDENT
UNIMATION, INC.

The Robot Institute of America is working with the U.S. Department of Commerce to consider the competitive posture of the USA in robotics. We shall be making recommendations as well as observations. By way of introduction I shall discuss the international scene.

1. In the industrial world (and even in some third world countries) there is an almost passionate fascination with robotics.

2. The Japanese hold an impressive edge in implementation. Their installed base is probably larger than that in the USA and Europe combined.

3. The levels of technology in the USA, Europe and Japan are approximately on a par.

149

4. The research frontiers in vision, tactile sensing, adaptivity are being addressed by academics and manufacturers throughout the world.

5. Governments in Japan, England, Germany and France have aggressive programs to support technology and to provide both internal and external market aids. This is also true in some of the Comecon countries. For example, development grants, government-owned application laboratories, licensing sponsorship, training grants, financing for users, trade barriers against imports.

6. Some examples of foreign government support:

a. Japan offers low cost rentals to small and medium size companies to induce them to adopt robot technology, at low risk.

b. USSR has taken a license under the most advanced German technology of KUKA, Keller and Knappich. Reputed to be a $50,000,000 deal involving hardware as well as technology. The controller is an advanced Siemens mini-computer with software from Berlin University.

c. French government is sponsoring robotics through funding manufacturers, users, academics and training schools. Reputed to be 430,000,000 francs.

d. The UK offers users financing for application studies by certified consulting firms.

e. The UK through the Department of Industry and the British Technology Group can support product development and production facilities. (Unimation Ltd. will start a new robot development and system facility with government sponsorship).

f. Using BIRD Foundation funds Israeli companies, in cooperation with U.S. companies, will develop advanced robotic peripherals.

g. Spain has a program to impose a 17 percent duty barrier on robots that compete with any produced by Spanish manufacturers.

h. The most advanced government sponsored robot application laboratory is the Institute for Production and Automation under Professor Warnecke of Stuttgart.

7. In the international marketplace the USA has some competitive products, i.e., Unimation, Cincinnati Milacron and Prab, but the exchange rate is a serious handicap to export.

8. The Cocom recommendations from the USA on control of robot export to the Comecon countries can only hurt the USA because the technology is in the public domain. Our foreign competition will serve this market.

9. The major U.S. companies entering the robot field are getting technology from abroad. IBM, GE, Westinghouse, United Technology and GM have all taken foreign licenses, primarily from Japan. The USA can legitimately fear the ultimate direct entry of these foreign companies, principally Japanese. Robotics is a "target industry" and U.S. companies in a "catch-up" mode are unwittingly seeding offshore company beachheads.

We would now like to dwell on two specific issues. Mr. Walter K. Weisel, incoming President of RIA and President, PRAB Robots, will discuss the need to enhance the U.S. robot market and Mr. Stanley J. Polcyn, President of RIA and Senior Vice President for Sales and Marketing, Unimation, will discuss industry views on industry-government relations.

The Need to Enhance the U.S. Robot Market

BY WALTER WEISEL
PRESIDENT
PRAB ROBOTS, INC.

I will try not to bury the importance of today's industrial robot's contribution to U.S. productivity in a sea of statistical data. Before presenting the industry data I want to discuss the basic technology. Today's industrial robot, in many cases, is over-qualified for the job that it is assigned and its technology closely parallels that of the NC machine tool business.

The major difference seen between NC machine tools and today's industrial robot is their method of application. Whereas only seventeen percent of all the manufacturing facilities in the United States currently use NC machines, it is forecast that virtually ninety percent of all industrial manufacturing processes for metal-cutting and metal parts transfer will use industrial robots by the year 1990.

ROBOTICS

An important difference in the robot's application base is that the robot has been applied in a wide range of non-metal producing industries such as textiles, construction, consumer goods, oil, medical and food.

Experience shows us that industrial robots applied to the various manufacturing processes in the industries mentioned has resulted in a gain of approximately twenty to thirty percent in productivity. This is a very important fact if you stop to consider that for a relatively small investment in an industrial robot a manufacturer can increase the output of his capital equipment in such a wide range of manufacturing and service industries.

It's important to know that the industrial robot is also a world processing tool for application in the industries mentioned above. Most industrialized countries are investing large sums of money into their industrial robot industry because of its large export potential and its ability to contribute to their sagging economy through increased productivity and reduced labor costs. If the United States is to become a dominant factor in the production and use of industrial robots the export market for the U.S. robot manufacturers is very important. It is estimated that in 1982 roughly twenty percent of the U.S. output of industrial robots was exported to either Europe or Japan.

The industrial robot was the invention of several U.S. companies and the first machines were installed here in 1961. By 1981, twenty years later, the total dollar volume of the U.S. robot industry had only reached $150 million. By the end of 1982 it was expected that the U.S. output of industrial robots would reach $200 million. Estimates of a thirty to thirty-five percent per year growth rate until 1990 are fairly common. It took twenty years to reach $150 million in sales. However, recent developments in increased labor costs, sagging productivity and the realization that robots are playing a major role in foreign countries has stimulated the need for industrial robots in the U.S. manufacturing community.

One of the major concerns of the U.S. robot manufacturers is the aggressive attitude of the Japanese for exporting equipment to the U.S. Many major U.S. corporations have signed marketing agreements with Japanese firms. They see their involvement with robots as a way of bolstering their current sagging sales and have rushed to embody the Japanese through manufacturing licenses and sales agreements in an effort to capture an early market share. U.S. corporations are rushing to put sales and service organizations in place to distribute the Japanese machines. Some industry experts estimate that by 1985 the Japanese will have twenty-five percent of the U.S. market, and I assure anyone reading this report that the Japanese won't wait until this is a $2 or $3 billion market before entering as they did in the case of the U.S. machine tool industry.

Japanese robot technology is not more advanced than that of U.S. manufacturers. The Japanese have been extremely aggressive in installing over 10,000 robots in their own plants. We believe that certain changes in our incentive system for puchasing U.S.-built robots are required in order to put this valuable technology into our factories as soon as possible.

If we don't move now to strengthen our own U.S. base of manufacturers, we will surely see unemployment rise as offshore manufacturers of robots displace U.S. workers in our industry. This has been the case in the U.S. auto industry where the massive use of industrial robots in Japan has made a major contribution to their reduced manufacturing costs for automobiles. The Japanese have increased their market share in the U.S. from five percent to twenty-five percent, thereby eroding the manufacturing base and creating unemployment. I don't mean to imply that the robot is the sole reason for the increase of Japanese imports, but it has played an important role.

It is our firm belief that the industrial robot of the 1980's will provide a base technology for implementing factory automation. Some experts have suggested that the industrial robot is the heart of the flexible manufacturing system that allows us to tie together a variety of key technologies. If, as a nation, we are to become a leader in factory automation, we must find ways to protect rudimentary technologies that are so vital to our success. We feel that the Department of Commerce has a golden opportunity to provide direction on how to keep an emerging industry before it becomes further threatened.

It would be a national tragedy if the very heart of our factory automation systems had to be provided primarily by offshore manufacturers. The U.S. manufactured industrial robot is worth paying close attention to and making sure it is implemented in a timely fashion.

Industry's Views on the Industry/Government Relationship

BY STANLEY J. POLCYN
PRESIDENT
ROBOTICS INSTITUTE OF AMERICA
SENIOR VICE PRESIDENT
UNIMATION, INC.

It has been identified that robotics is a market targeted by foreign nations, and that the U.S. domestic market is weak, with a growth rate that lags Japan and Europe. The following are some courses of action to enable us to react to foreign targeting and also to develop and grow our domestic and export market.

Other nations believe that they have a sovereign right to do whatever they wish to, be it proper or improper, to protect their industries and markets and to pursue ours. We know the approach and have models of the way this is done:

A. Protect domestic markets.
B. Buy into export markets with low prices.
C. Use cheap money for financing and consignments.

D. Take a joint venture approach or licensing to obtain technology and/or markets.

We know we cannot change them but we can change us. Our government must bear the responsibility for standing up against those nations that are doing targeting and take a very positive position and draw a very firm line.

We believe our government, led by the Department of Commerce, could:

A. Establish robotics as a major strategic industry;

B. Set national productivity goals that will enhance our industrial base;

C. Establish an Industrial Policy Board (similar to the Federal Reserve Board) that can take forceful, effective and independent action to strengthen its strategic industries.

1. The Board should be a small working group and backed by strong organizations.

2. It should be a coalition for planning among manufacturers and be absolved from any anti-trust regulations.

3. Its charter should include power to take fast action on its resolutions and be reactive to external threats as they develop.

D. Provide an aggressive, consistent taxation policy that would provide incentives to utilize robotics technology.

1. Restructure depreciation schedules to reflect real world technical obsolescence and provide incentives to those taking innovative risks.

a. Provide a 5-year double declining for U.S. manufactured robots only. Straight line for foreign.

2. Establish 25 percent investment tax credits for U.S. manufactured robots—no ITC for foreign.

3. Provide special tax credit to users for first time applications that are innovative to compensate for risks taken.

4. Provide special tax credits to users of robots in hazardous environments such as asbestos manufacturing or munitions.

5. Provide special tax credits to robot manufacturers for those robots that are exported to enable them to compete in export markets that have duty barriers, unfavorable exchange rates or governmental regulations that provide barriers to their market.

6. Provide special tax credits to those doing robot research and development, both basic and applied.

E. Create and fund a robot leasing company that would provide low-cost loans to purchasers of industrial robots manufactured domestically to encourage robot installations and increasing production.

1. Provide special low-rate loans:
   a. to small companies employing less than 250 people;
   b. for installations in dangerous jobs or environments;
   c. for first time innovative applications of robots.
2. Provide special loans to foreign buyers of robots to support major export programs of domestic robots or robotic systems.

F. Provide government funds to assist in generic, basic and applied research for advancing robotic developments so that the industry maintains its leadership in the future.

1. Permit companies in the industry to get together without fear of anti-trust action, to plan and assign development projects to avoid multiple duplication of effort.
2. Pass University research funds through robot manufacturers or the Robot Institute of America to obtain a more efficient and directed total robotic research program.

G. Work with the Department of Commerce and other components of government to promote a realistic export program. The Department of Commerce should continue to permit export of robotic products under General License. Manufacturing technology and software source codes could be restricted from export to Comecon countries if necessary to meet national security objectives.

It is necessary that we eliminate any adversary condition between government and industry and set a political climate to encourage cooperation toward national goals. It is time that we step up and protect the legitimate interests of the United States. The future of our country depends on a healthy growing industrial base. The issues presented relate to increasing that industrial base and are above the issues of partisan politics.

Summary of Specific Issues Raised by the Robotics Industry

Summarizing the current issues of particular concern to the robotics industry:

■ Other countries' targeting practices are seen as a major concern because they put U.S. manufacturers at an unfair competitive advantage in both domestic and international markets.

■ The sluggishness of the U.S. domestic market was cited as a major problem. The industry suggested various policy initiatives and fiscal and tax measures to provide incentives for U.S. firms to purchase robots.

■ Antitrust law and regulation was cited as a hindrance to efficient R&D because of the possibility of governmental and private suits for cooperative R&D.

161

■ Export controls were a concern because of their perceived potential for hindering U.S. manufacturers more than foreign suppliers in their efforts to sell abroad.

■ The industry experts noted that other governments provide extensive support for R&D and technical training as a targeting technique. The experts believed that greater government investment was desirable in R&D for robotics applications.

4
The Telecommunications Industry

A Study of the International Competitive Position of the U.S. Telecommunications Equipment Industry

BY ROBERT ECKELMANN
U.S. DEPARTMENT OF COMMERCE

Profile and Summary

Over the past decade, the U.S. telecommunications equipment industry has enjoyed the solid growth (6-8% per year in real terms) typical of a dynamic high-technology sector. Total domestic American production recently passed the $20 billion mark, and despite a current lull, most analysts expect that the sector will soon resume its healthy advance and prove capable of sustaining a 5-9% annual pace for at least the next 5-10 years.

From an international point of view, the 1980s represent a critical period for U.S. manufacturers, who at present account for over 40% of world production. As in other information industries, rapid technological advances will make telecommunications an especially fast-moving area, and continued success will require allocation to research and development of the considerable resources needed to stay at the frontiers of innovation. At the same time, traditional methods of industrial targeting, and the foreign competition they promote, will further intensify the challenge facing American firms. But also at work are additional factors unique to telecommunications, most notably the dramatic changes now underway in the marketplace.

Traditional service monopolies in certain countries are now being restructured to encourage competition and improve overall efficiency. But other major markets remain essentially closed, operating according to the established system of government-held PTTs and preferential procurement procedures. As long as these two systems coexist among the major producer countries, as long as this type of asymmetry persists, firms based in open-market countries (and therefore without the luxury of insulated domestic demand) could find themselves at a temporary competitive disadvantage before their protected counterparts. And since the U.S. has led the way in terms of liberalization, American telecommunications equipment manufacturers in particular may, until a more balanced situation develops, face unreciprocated foreign competition both at home and in open markets abroad.

Once again, our most formidable rival will be Japan, which currently enjoys an annual trade surplus with the U.S. in telecommunications equipment of nearly $250 million (1981). Its leading companies have demonstrated exceptional technological capabilities and broad marketing strength. Selected European firms will also challenge the American leaders—Ericsson of Sweden and Philips of the Netherlands have already proven their ability to compete effectively without the support of sizeable captive markets.

But most important for the future international competitiveness of the U.S. telecom equipment industry is improved access to foreign markets. Lately, there have been a few, but only a few, positive signs—some monopoly dissolution in Canada and Britain, and limited bilateral procurement agreements such as that struck with Japan on the issue of NTT. The telecommunications field plays a central role in the current high-technology revolution; it is emerging as a centerpiece of the post-industrial economy. But the future well-being of its supplier industries will depend critically upon their ability to perform in the international marketplace. At present, foreign restrictions seriously constrain this essential type of growth and development. It is imperative that the U.S. government make full use of both bilateral channels and multilateral frameworks to minimize such interference, with the objective of affording American producers equitable opportunities for competition and growth.

Definition and Scope

The term "telecommunications", though still in many ways a convenient label, no longer specifies a discrete high-technology sector. A generation ago, virtually all equipment used to transmit, carry, and receive information was included in the central telecommunications network and regulated by the Federal Communications Commission (FCC). But competition and deregulation, together with changing technology, have now revolutionized the industry. Today, specialized carrier switches and microwave links, domestic satellite company two-way earth stations, and packet-switching exchanges are all superimposed upon the traditional common-carrier core network. In addition, the proliferation of data processing systems and equipment has blurred traditional distinctions between telecommunications and fields ranging from computer software to aerospace. But for analytical purposes, a few key restrictions can help to section off the central areas of the industry in ways that, while conceptually unsatisfactory, do enable detailed study and avoid unnecessary overlap with related sectors. These limitations narrow the focus of this profile to three general categories of telecommunictions equipment:

    1. switching equipment, which serves to connect terminals and to coordinate entire telecommunications networks from central offices;

    2. terminals (from simple telephones to digital PBX systems, from teletypewriters to facsimile equipment), which transmit information through the network from one end user to another;

3. transmission equipment (from simple paired wire cable to optical fibers, from coaxial cable to microwave).

In general, this study will *exclude* equipment used to manipulate information in any way except what is needed for transmission (e.g., data processing equipment). It will also avoid, where possible, tangentially related areas of consumer electronics, such as CB/transistor radios and home television sets.

Even so, definitional problems still loom large for any analysis of the telecommunications sector. International discrepancies, inconsistent methodologies, and, on occasion, a sheer lack of reliable data require that this profile invoke several product classification systems and codes in the course of its discussions:

1. The narrowest of these, Standard Industrial Classification #3661 ("Telephone and Telegraph Equipment"), will be applied in most analyses of domestic output. In certain cases, it will by necessity be combined with segments of SIC 3662 ("Radio & Television Communication Equipment") to include the mobile radio, fiber optics, facsimile and satellite equipment components of the industry.

2. The TSUSA and Schedule B classifications will be used in determining the value of certain U.S. trade flows. These systems do not match up perfectly with the SIC 3661 heading for domestic production. The trade data includes, for example, radio transmitters (from SIC 3662), while excluding such output as microphones.

Further clarification will accompany any data presentation where inadequate or incompatible information has necessitated an adjustment of this classification framework.

Finally, this report does not address the area of telecommunications services, except in noting that suppliers in that market are in turn the primary consumers of telecommunications equipment. The monopoly power of these publicly controlled entities has had important consequences for the producers from whom it procures; and now deregulation, intensifying competition, and technological change have begun to transform the shape of this once compact and well-defined industry. The future direction and ramifications of these trends will receive due consideration as key issues for telecommunications equipment manufacturers. But as developments in their own right, they will remain at the fringes of our analysis.

The U.S. Market— Consumers and Producers

Over the last ten years, the U.S. telecommunications equipment sector (SIC 3661) has expanded steadily. A real compound annual growth rate of 8.1% during that period culminated in total 1981 shipments of nearly $12.2 billion. And although expectations for the industry's 1982-83 performance have moderated somewhat (1983 estimates include only a 2.1% real rise to $14.5 billion), most analysts predict that the solid pace of the 1970s (5-6% per year) will resume soon thereafter.[1]

The most prominent American "consumer" of telecommunications equipment is AT&T, responsible for some 58% of total U.S. spending and, through its Bell System, servicer for approximately 80% of the national telephone network. Independent telecommunications companies accounted for an additional 18% of U.S. equipment demand in 1982. Finally, interconnect companies (focusing on "private" communications systems), specialized common carriers (such as MCI, Southern Pacific, etc.), domestic satellite firms, and value-added carriers constituted the most dynamic group of telecom equipment purchasers, expected to provide nearly 25% of demand in 1983 (a 43% value increase over 1982 figures).[2]

The production side of the U.S. telecommunications equipment industry features four leading firms, accounting collectively for around 90% of domestic supply. The major role here belongs to Western Electric, whose 1982 output of $13.0 billion meant a 68.4% market share.[3] The other three leading firms—GTE, North-

1. U.S. Industrial Outlook, 1983.
2. Telephony, January 1983.
3. Western Electric output figures reflect equipment value upon installation. A.D. Little's compensated estimates for 1982 W.E. production ran as low as $8.9 billion.

Table 31
The U.S. Domestic Telecommunications Market

Equipment Demand	1982 Estimates	1983 Forecasts
Bell Telephone Companies*	$17.0 billion	$16.8 billion
Independent Telecommunication Companies	$ 4.8 billion	$ 5.1 billion
1. General Telephone & Electronics (GTE)		
2. United Telephone Systems		
3. Continental Telephone		
4. Central and Utility Telephone		
5. Mid-Continent Telephone		
Interconnect Companies and Affiliates	$ 4.0 billion	$ 5.5 billion
Specialized Common Carriers	$ 0.9 billion	$ 1.5 billion
TOTALS	$26.7 billion	$28.9 billion
of which SIC 3661 comprised	$14.5 billion	
telecom portions of SIC 3662 comprised	$ 4.4 billion	

Equipment Supply (SIC 3661/partial SIC 3662)	1982 Estimates	Market Share
Western Electric (AT&T)	$13.0 billion	68.4%
GTE	$ 2.1 billion	11.1%
Northern Telecom	$ 1.5 billion	7.9%
ITT	$ 0.6 billion	3.2%
Others	$ 1.8 billion	9.5%
TOTALS	$19.0 billion	100.0%

Sources: Telephony; Bureau of Industrial Economics estimates; Corporate Annual Reports.
*Includes AT&T, its principal telephone subsidiaries, and Bell affiliates.

ern Telecommunications, and ITT—together comprised an additional 22.1% of total U.S. production activity. The remaining 9.5% derived from either imports or smaller domestic manufacturers.[4]

Finally, this high-technology field's steady efficiency gains have been reflected in recent industry trends for both price and employment. As shown in Figure 4, the telecom equipment price deflator has, over the last several years, consistently trailed its counterpart for the economy as a whole; at the same time, because of productivity improvements, consistent growth in output has coincided with a decline in hiring. And most analysts expect both trends to continue through much of the 1980s.[5]

The World Market

Global production of telecommunications equipment in 1981 was valued at approximately $45 billion, of which $9 billion was com-

Figure 4
Price & Employment Trends in Telecom Equipment

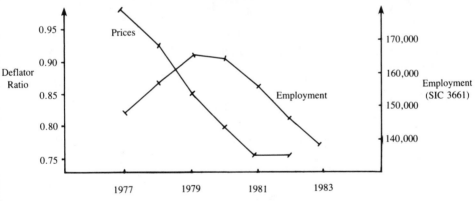

Sources: U.S. Industrial Outlook, 1983; Survey of Current Business.
Note: Deflator Ratio is calculated by dividing the GNP deflator into the deflator for telecommunications equipment.

4. Bureau of Industrial Economics Estimates.
5. Source: U.S. Industrial Outlook, 1983; Survey of Current Business.

mercial output covered by a relevant portions of SIC 3662.[6] Tables 32 and 33 provide a breakdown of these industry totals both by country and by firm.

But while over 90% of world output still emanated from the advanced industrialized countries, demand has evolved according to a more diversified pattern, with Europe and the developing countries expected to show the most rapid growth from 1977 to 1987 (see Figure 5 for an illustration of these trends). The reasons behind these changes range from market liberalization to the need for large-scale systems projects; but above all, such shifts will accentuate the importance of both telecommunications trade and multinational operations for continued industry growth.

A breakdown of the basic product lines in the telecommunica-

Table 32
The World's Leading Telecommunications Equipment Producers
(value, in bil $)

By Country	1981	By Firm	1976	1981
United States	21.8	Western Electric	6.93	11.53
Japan	5.6	ITT	3.35	5.50
Europe	17.0	Siemens	1.96	4.40
of which France	4.8	Ericsson	1.56	2.48
W.Ger.	3.9	GTE	1.33	2.23
U.K.	2.4	NEC	0.84	2.05
Italy	1.4	Northern Telecom	1.08	1.88
Sweden	1.2	Motorola	0.95	1.60
Canada	2.2	Thomson	0.60	1.45
TOTAL(est)	45.6	G.E.C.	0.53	1.45
		Philips	0.82	1.30
		C.G.E.	0.63	1.00
		Plessey	0.39	0.81

Sources: MacKintosh; A.D. Little; Telephone Engineer and Management; Bureau of Industrial Economics.

6. See "Definition and Scope" section for more detailed clarification and discussion.

tions field and a summary of their respective shares of the total equipment market are provided below in Table 33. The most recent data available comes from 1975, but asterisks mark the high-growth segments of the market for whom observed values could have changed appreciably.

Telecommunications Trade

Over the last decade, all major producer countries have witnessed more rapid growth of their telecommunications exports and imports

Table 33
Product Line Shares of World Sales

		Percent of Total
Switching		33.6
of which Central Office	24.8	
Private Branch Exchanges/PABX	4.6	
Message Switching and		
Communications Processors	4.2*	
Transmission		32.8
of which Microwave radio and MUX	5.7*	
Longhand and coaxial cable	5.0	
Local wire and cable	21.4	
Satellite and earth stations	0.8*	
Terminals		13.0
of which Telephone subscriber sets	8.4	
Data Communications Terminals	3.4*	
Modems	0.8*	
Teletypewriters	0.4	
Mobile Radios		7.6
Private Systems		8.4
Other		4.6
TOTAL		100.0

Source: OECD, 1982. See text above for explanation of asterisks.
Note: Many such surveys exclude cables, lending greater importance to switching and terminal equipment while diminishing the role of the transmission segment of the market.
*High-growth market segments, where appreciable change may have occurred since 1975

Figure 5
Regional Shares of the World Telecom
Market—1977, 1982, & 1987
(expressed in 1982 $)

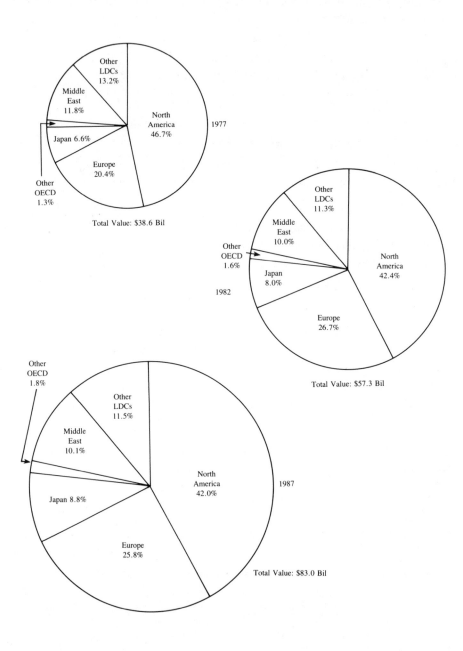

Table 34
World Trade in Telecommunications Equipment (SIC 3661 only)

Principal Producer Countries	1977 Imports	Exports	(in million $) Balance	1981 Imports	Exports	Balance
Japan	24	363	+ 339	46	911	+ 865
Sweden	36	458	+ 422	65	776	+ 711
West Germany	104	562	+ 458	128	809	+ 681
Netherlands	126	228	+ 102	128	398	+ 270
France	57	168	+ 111	86	320	+ 234
United States	129	257	+ 128	494	653	+ 159
Canada	93	80	− 13	143	298	+ 155
United Kingdom	91	247	+ 156	235	331	+ 96
Belgium/Lux	76	248	+ 172	118	262	+ 144
Italy	50	97	+ 47	101	143	+ 42
TOTALS	786	2708	+ 1922	1544	4901	+ 3357

than of their domestic markets. This has led to a gradual reversal of the traditional orientation of equipment manufacturers towards demand from national service monopolies. Table 34 demonstrates this development and summarizes the current trade position of the principal supplier nations.

U.S. Trade

Over the last five years, the American position in telecommunications trade has weakened perceptibly. Though exports have grown at a fairly consistent pace (13-18% per year), a continuing surge in imports (24-30% per year) has caused stagnation, and most recently deterioration, of the U.S. surplus. Because of the "major project" nature of much telecommunications trade, local year-to-year figures often prove volatile and misleading, but Table 35 portrays the aggregate trends that characterize recent American progress.

Depending upon one's definition of the sector, the 1982 trade balance ranges from + $90 million (SIC 3661 only) to + $450 million (see Table 36). In either case, this represents a 35% drop since 1980

Table 35
Aggregate Trends in U.S. Telecom Equipment Trade
(SIC 3661 only, mil $)

	1972	1977	1979	1980	1981	1982	1983	1977–83 growth rate
Exports	76	257	448	557	653	725	850	+22.1%
Imports	86	129	319	421	494	635	790	+35.3%
Balance	−10	+128	+128	+136	+159	+90	+60	−11.1%

Source: 1983 U.S. Industrial Outlook.

alone, with an additional decline of 20-33% forecast for 1983. Japan accounted for nearly 50% of U.S. telecommunication imports, resulting in a bilateral deficit approaching $250 million. Figures 6 & 7 provide an overview of the sources of U.S. imports, the destinations of U.S. exports, and the trade flows between the U.S. and its major competitor nations in the telecommunications equipment field.

Japan

The Japanese telecommunications equipment industry poses the greatest national competitive challenge to American manufacturers. As in several other high-technology fields, Japan has succeeded in fostering, through public policy promotion, the rise of a highly efficient and technologically sophisticated telecommunications capability. As in the cases of computers, robotics, and semiconductors, this was accomplished first by achieving control over a national market, then by moving agressively into the international picture through trade and foreign investment.

Although dozens of firms share in the Japanese telecommunications market, four companies dominate in terms of both production and sales: NEC, Oki, Fujitsu, and Hitachi. Table 36 indicates their size, relative position, and recent rates of growth.

As indicated earlier, the demand faced by these manufacturers has grown vigorously and consistently over the last decade, fueling

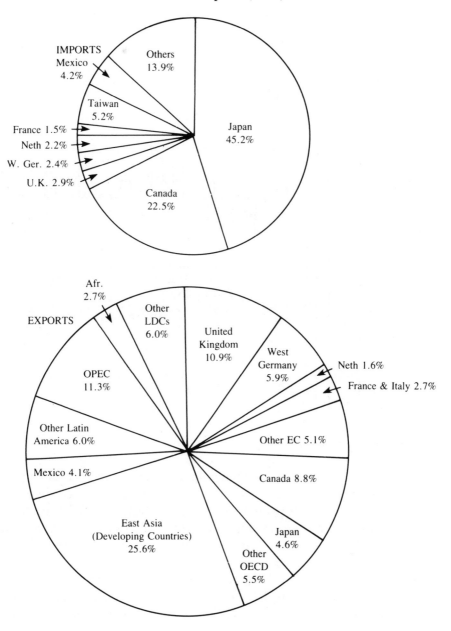

Figure 6
Sources of U.S. Imports (1982) and
Destinations of U.S. Exports (1982)

IMPORTS

Mexico 4.2%
Others 13.9%
Taiwan 5.2%
France 1.5%
Neth 2.2%
W. Ger. 2.4%
U.K. 2.9%
Japan 45.2%
Canada 22.5%

EXPORTS

Afr. 2.7%
Other LDCs 6.0%
United Kingdom 10.9%
West Germany 5.9%
Neth 1.6%
France & Italy 2.7%
OPEC 11.3%
Other EC 5.1%
Other Latin America 6.0%
Mexico 4.1%
Canada 8.8%
East Asia (Developing Countries) 25.6%
Other OECD 5.5%
Japan 4.6%

Source: Official Trade Publications

Figure 7
U.S. Bilateral Trade Position in Telecommunications, with Selected Countries (1981, mil $)

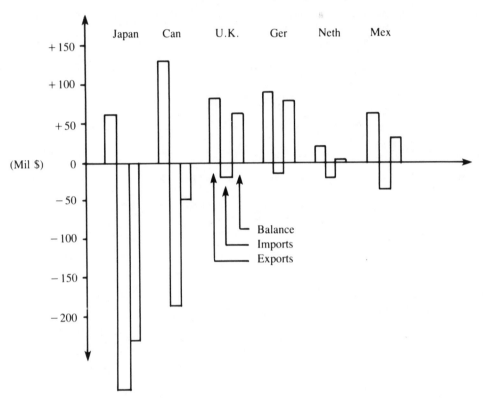

Source: Official U.S. Trade Publications.

their sustained expansion. But most important for their development as a secure and healthy industry was the protection afforded by their main customer, Nippon Telephone and Telegraph (NTT), through its closed procurement procedures. Sheltered from the challenge of their foreign peers, but engaged in competition among themselves, the Japanese were able to promote the steady advance of a sector whose importance to their high-technology future was never underestimated. Through the 1970s, the Japanese telecommunications equipment market grew at a hearty 6.8% per annum rate, approach-

178

Table 36
Japan's Leading Telecommunications Equipment Suppliers (1981)

Firm	Domestic Sales (Mil $)	Market Shares
Nippon Electric Company	1,620	36.9%
Oki Electric Industry Company	262	6.0%
Fujitsu	567	12.9%
Hitachi & Hitachi Denshi	497(Est)	11.3%
Sub Total	2,946	67.1%
TOTAL Japanese Market	4,486	100.0%

Source: Pacific Projects, Ltd. Exchange rate used: 215 Y/$.
Note: Excludes export production, valued at $1.24 billion in 1981.

ing the $5 billion mark in 1982. And forecasters expect the 1980s to bring continued rapid expansion (7.5-8.0% per year in real terms), meaning annual sales of nearly $9 billion by 1986. Buoyed by this growth, the nation's industry has emerged as an international force to be reckoned with. Japan's telecommunications trade figures, presented in Tables 37 and 38, reflect both the impenetrability of its domestic market and the steady advance of its industry into a full range of overseas opportunities.

Recent developments indicate that some change is underway. As part of a 1981 accord between the U.S. and Japan, NTT agreed to

Table 37
Japanese Local Production and Trade by Segment
(billion $)

Category	Local Output		Imports		Exports	
	1981	1986	1981	1986	1981	1986
Telephone & Telex Equipment	2.55	4.23	0.01	0.02	0.52	1.36
Transmission Equipment	2.12	2.98	0.06	0.10	0.43	0.69
Mobile Radio	0.41	0.69	—	0.06	0.10	0.16
Video & Radio Broadcasting	0.43	0.73	—	0.01	0.18	0.33
Data Communications	0.14	0.25	0.01	0.02	0.02	0.03
TOTAL	5.64	8.88	0.08	0.21	1.24	2.56

Table 38
Japanese Telecom Trade (mil $)

	1978	1981	1986	Growth Rate
BALANCE	+621	+1,120	+2,353	18.0%
Imports	113	81	209	7.2%
Exports (Total)	734	1,241	2,562	17.1%
as % of total production:	17.1	22.0	28.8	

Source (both Tables): Pacific Projects, Ltd.

revise its procurement policies for telecom equipment to enable
more frequent and more equitable participation of foreign firms.
While some American sales have resulted from the restructuring, its
overall impact thus far has been quite limited. Time may make a difference as both sides adapt to the opportunities presented by their
new relationship. But the sluggish pace of change to date has for
many cast doubts on the eventual effectiveness of this bilateral
agreement, either as a mechanism for opening the Japanese marketplace or as a precedent for resolving similar procurement problems
in other major countries.

The Japanese Government's Approach to Telecommunications

The Japanese Government's promotion of telecommunications development has followed many of the same fundamental principles
applied in its approach to other high-technology sectors. The concepts of managed growth, vigorous but continued competition, and
domestic success as the foundation for a subsequent international
push—these anchor the philosophy which has guided Japan's industrial policy treatment of several major industries.

However, because of the unique features of the telecommunications market, the application of traditional ideas took some unique
and unprecedented forms. The main mechanism invoked for the

purpose of encouraging Japan's progress in telecommunications was the national service monopoly, Nippon Telephone and Telegraph (NTT). Under its beneficient umbrella, Japanese equipment manufacturers engaged in controlled competition, sheltered above all from the challenge of foreign producers. Some supportive measures did provide direct incentives (special tax treatment, access to participation in large-scale R&D) for firms to become more active, efficient, and technologically advanced suppliers. At the same time, NTT encouraged both competitive vying for prominence within its supplier network and the gradual absorption of technology from abroad. The eventual result, achieved via methods that have imposed formidable costs upon the monopoly and its customers, has been the emergence of an internationally competitive sector, now expanding rapidly into overseas opportunities and ostensibly willing to open segments of its market at home.

The critical question that remains is what form the Japanese government's support of future telecom development will take. The international controversy and domestic efficiencies surrounding NTT's procurement methods, the gathering strength and capability of Japanese telecom firms, the growth of major R&D projects aimed at both autonomous advancement and the integration of technologies from related industries—these factors may point to possible shifts in Japanese methods for telecommunications development. Above all, this shift could include a gradual decline in the role of NTT—and a corresponding (though passive) increase in the importance of conventional promotion methods—in industrial policy efforts at work in this particular field.

Western Europe

The European telecom market has expanded at the rapid rates observed in other industrialized areas, and, depending upon the type and degree of market liberalization that occurs, future performance could be equally impressive. Overall, diversified demand (lodged overwhelmingly with national PTTs) is met, as elsewhere, by a rela-

Table 39
The Leading European Telecommunications Equipment Firms

Company	Telecom Sales (Bil$)		Growth Rates
	1976	1981	
Siemens (W. Germany)	1.95	4.40	17.6%
Ericsson (Sweden)	1.56	2.48	9.7%
Thomson (France)	0.60	1.47	19.6%
G.E.C. (United Kingdom)	0.53	1.46	22.5%
Philips (Netherlands)	0.82	1.28	9.3%
C.G.E. (France)	0.63	1.00	9.7%
Plessey (United Kingdom)	0.39	0.80	15.5%

Sources: OECD, ADLittle.

tively small group of large equipment manufacturers. The next two tables provide a listing of these leading European companies, followed by summaries of national production and consumption figures.

The telecommunications equipment market in most European countries falls into one of these categories:

1. essentially closed, with supply dominated by one or two large, domestically-oriented manufacturers (France, Germany);

2. in the process of liberalization, and therefore showing less concentrated national production, greater foreign penetration, and (stemming from a more competitive industry) some export of foreign subsidiary capabilities (United Kingdom);

3. smaller and fairly open, either home to a major export oriented firm (Sweden, the Netherlands) or reliant upon foreign sources for their equipment supply (Austria, Ireland).[7]

7. As background to the British Telecommunication Act, Logica Ltd. developed a "liberality" matrix to characterize national European telecom markets in each of four categories: resale of facilities to third parties, connection of privately supplied equipment to common carrier facilities, regulations governing the use of intraorganization private circuit networks, and network charges (tariffs) made for private circuits. Countries were ranked in each category on a scale of 1 to 3 (higher numbers meaning more liberal treatment): a "3" means generally permissible, few restrictions, or, for tariffs, low cost charges; a "2" means

TELECOMMUNICATIONS

Of course, the rate and impact of future liberalization is difficult to predict; but should developments continue along their current course—should the basic pattern set in Britain, Canada, and Australia spread to include other key markets—the prospects for the continent's telecom progress could become bright indeed. The U.S. Government is now actively pursuing a multilateral trade agreement for interconnect equipment. At this point, however, the European situation presents at best a checkerboard of opportunities and problems in the telecommunications field.

As a single entity, the European telecom market is nearly the size of its U.S. counterpart. But due again in part to the lag in liberalization (and hence the imbalance between demand for imports at home and abroad), the continent enjoys an impressive collective surplus in telecommunications equipment trade.

The aggressive export orientation of selected firms (Ericsson and Philips in particular, both with approximately 80% of their sales coming from abroad) has also contributed heavily to this trend. But on an overall level, the pattern primarily indicates the major role European manufacturers still play in third markets in determining the international competitive position of the American telecom equipment industry.

permitted with restrictions, or, for tariffs, moderate charges with some punitive element; a "1" means permitted only to governmental authorities, or with heavy restrictions, or complete prohibition, or in the case of tariffs, punitive charges designed to discourage.

Country	Resale	Connect'n	Regulation	Tariffs
W. Ger.	2	1	1	1
Switz	2	1	1	1
Spain	2	1	1	2
Italy	2	1	2	2
France	2	2	2	2
Belgium	2	2	2	2
Sweden	2	1	2	3
Neth	2	2	2	3
U.K.	2	2	2	3
U.S.	3	3	3	3

Table 40
Telecommunications Equipment in Europe

Production	1978	1981 (est)	Growth Rate	Share
France	3,113	5,457	20.6%	30.3%
West Germany	2,607	3,919	14.6%	21.8%
United Kingdom	1,552	2,439	16.3%	13.5%
Italy	1,150	1,461	8.3%	8.1%
Sweden	885	1,229	11.6%	6.8%
Netherlands	625	1,059	19.2%	5.9%
Belgium	700	1,011	13.0%	5.6%
Others	1,069	1,426	10.1%	7.9%
TOTAL	11,701	18,001	15.4%	100.0%

Consumption	1981	1984	Growth Rate
France	4,383	5,679	9.0%
West Germany	3,334	4,001	6.3%
United Kingdom	2,302	3,066	10.0%
Italy	1,389	1,806	9.1%
Sweden	374	377	0.3%
Netherlands	609	658	2.6%
Belgium	816	945	5.0%
Others	1,495	1,747	5.3%
TOTAL	14,702	18,279	7.5%

Source: MacKintosh Publications, Ltd., 1981.

Other Key Issues in the World
Telecommunications Market

Liberalization in the Services and Equipment Markets

Effects on Supply and Demand
Many of the most notable events in the telecommunications world
have involved fundamental changes, of thought and practice, in the

Table 41
Trade and Export Performance of Major European
Producer Nations (mil $)

	Trade Balance	Exports	Export Growth Rate	Shares of World Exports*
West Germany	+ 681	809	10.0%	17%
Sweden	+ 711	776	14.0%	16%
Netherlands	+ 270	398	15.0%	8%
United Kingdom	+ 96	331	8.0%	7%
France	+ 234	320	17.0%	6%
Belgium/Luxembourg	+ 144	262	1.0%	5%
Italy	+ 42	143	10.0%	3%
TOTAL	+2,178	3,039	10.8%	62%
(United States	+ 128	653	26.3%	13%)
(Canada	+ 155	298	38.0%	6%)
(Japan	+ 865	911	25.9%	19%)

Sources: U.S. Industrial Outlook; Official Trade Publications.
*Includes the countries listed, Japan, the U.S., and Canada—a group that accounts for roughly 90% of total world exports. Figures are for 1981, SIC 3661 only. Growth rates are calculated for 1977–1981 period.

structure of major national markets. Though telecom service was long considered a textbook case of a justifiable monopoly (on the grounds of economic efficiency), a combination of expanding functions and steady technological change gradually reordered traditional thinking in ways that now emphasize the competitive aspects of the industry. Due to the unprecedented nature of these developments, no one can predict their full practical implications with confidence. Nevertheless, the theoretical aspects of recent changes may offer some interesting clues.

In a pure sense, dissolution of an established service monopsony (such as AT&T) would present competitive equipment manufacturers with both greater demand and higher product prices. However, should market liberalization also facilitate entry on the part of equipment suppliers for structural reasons[8]—either because of previous exclusionary procurement practices, other types of formal barriers (such as technical standards/requirements), or the opening of international flows—prices could again be pressured downward. Regardless of which particular conditions apply, telecommunications services would become more plentiful at lower cost, with both the supply of and demand for equipment rising as well. The final result: a more dynamic and efficient industry structure.

International Implications

"Liberalization" also gives new meaning to the concept of a world market in telecommunications (given the previous practice of "national" procurement and supply). An open international environment would surely result in rapid growth in telecom trade—more intense foreign, as well as domestic, competition.[9] And under fully competitive conditions, a global position could enable realization of the full economies of scale available in the equipment market.

But the current situation falls somewhat short of that just described. Few countries have acted out of a commitment to liberalization, for predictable sorts of reasons:

1. their simple resistance to change slows implementation;

2. depending upon the international competitiveness of their domestic equipment industry, the maintenance of a closed, captive market may be utility-maximizing;

3. the primary beneficiaries of liberalization (consumers) are the most scattered and diversified of interest groups; while,

4. the victims of reform (monopoly servicers and the non-competitive suppliers they shelter) can more easily be molded into a cohesive, influential group.

As a result, the present world telecom equipment market is characterized above all by *asymmetry*, with certain countries having

8. vs. price reasons. The price rise mentioned in the previous sentence could not be neutralized by new market entry for price reasons alone.

9. There is already significant evidence of such a trend in national markets where significant liberalization has taken place. The value of U.S. telecom equipment trade has grown from $386 million to $1,360 million in only 5 years (1977-1982, SIC 3661 only), a 29% annual rate of increase.

opened both the equipment and service areas, with others having liberalized but one of those segments, and with still others essentially closed altogether. And across all three cases, change is progressing at vastly different rates.

One direct consequence is an otherwise puzzling imbalance in telecommunication equipment trade. For the most liberal nations (a group decidedly in the minority), open demand at home attracts foreign competition, while closed markets abroad mean few opportunities for export to *competitor* countries. Ironically, the less efficient producer countries, because of market restrictions at home, often enjoy the greatest trade surplus.

For the U.S., openness in an industrialized world that is otherwise largely protected has meant that, despite being the world's technological leader and the strongest performer under competitive conditions at home,

1. it enjoys the smallest trade surplus (as a percentage of production) of any major OECD country;

2. it suffers a net deficit with those same trading partners in telecommunications equipment;

3. over ½ of U.S. exports go to developing countries, and over ¾ of U.S. exports go to those third markets plus Canada and the United Kingdom (the most liberal industrialized markets).

Only continued liberalization will remedy this imbalanced, asymmetric situation. Meanwhile, foreign manufacturers who can launch their overseas efforts from captive markets at home could enjoy a structural competitive advantage over their deregulated counterparts.

Competitive Consequences

Because of the manner in which these differing degrees of liberalization skew the world telecommunications market, international competitiveness in the industry becomes extremely difficult to assess and measure. Many analysts argue simply that those manufacturers who operate and succeed under the most open conditions are the most efficient, and hence the most competitive, suppliers. Others place greater emphasis upon the support a captive market can provide for producers who then also launch out into an unprotected market.

There is, no doubt, an element of truth in both lines of reasoning. But they still leave unaddressed the issue of competitive strength. Because of extensive international impediments, conventional trade figures fail to reflect meaningfully the main supplier countries' relative positions. Relative performance in third markets provides a seemingly attractive alternative; but because of both tremendous variations in individual figures (see next section) and traditional procurement/bidding irregularities, the available data again proves generally unsatisfactory.

But a simple comparison of import penetration levels in the major markets, though still an imperfect approach, offers a neutral gauge which can then be interpreted in light of each nation's apparent restrictiveness. Table 42 presents the most recent of these statistics (1981) for SIC 3661: —

The data gives a reasonable indication of the competitive strength of the U.S. telecommunications equipment industry:

1. The numbers on their own show that import penetration of the U.S. market is less than the average for the major supplier countries, with only the Japanese and French producers enjoying a greater share of their respective domestic markets. In most cases, the fractions involved, and the disparities between them, remain fairly small.

2. U.S. capabilities become more impressive when one considers that domestic manufacturers turned in their above-average performance despite being situated in the most open of the world's major markets (i.e., despite facing the full force of international competition).

3. Finally, Table 42 may further understate U.S. success at home and abroad since, by considering only SIC 3661 equipment, it excludes many of the highest-technology areas in which the American industry particularly excels (satellites, radar, etc).

Procurement requirements and restrictions can and do seriously impede foreign efforts to market telecommunications equipment in several countries. The above figures on import penetration point to where such constraints may play the greatest role. But the statistics

Table 42
Import Penetration of Major Suppliers' Home Markets
(as % of production and consumption)

	as % of Consumption	as % of Production
United States	4.1%	4.0%
Japan	1.1%	0.9%
Canada	7.0%	6.4%
Europe*	11.3%	9.0%
France	2.0%	1.9%
West Germany	4.1%	3.3%
Italy	7.4%	7.1%
United Kingdom	10.4%	9.9%
Sweden	14.0%	5.8%
Belgium	14.5%	11.7%
Netherlands	21.0%	12.1%

Source: Official Trade Publications, MacKintosh (*op. cit.*), and the U.S. Industrial Outlook.
*Data is for 1981, SIC 3661 only; "Europe" includes only those countries listed (i.e., those countries whose import penetration levels are lowest).

also convey the enduring competitiveness of American producers in spite of international asymmetries, indicating the enormous potential value and significance of improving U.S. access to foreign opportunities and demand.

Technological Effects

Intensified competition between U.S. suppliers of telecom equipment should contribute to improvement of the industry's technological capabilities. More open procurement and production environments should impress more heavily the need to enhance product sophistication and performance. Liberalized conditions should speed the already impressive advance of telecommunications technology.

But at the same time, that openness could serve to narrow what innovation "gap" persists between the U.S. and its overseas telecom rivals. As the latter strive to compete in an accessible American system, they will be forced as well to move to the frontiers of research and development. The internationalization of the demand they face will accelerate their technological progress, a stimulus that their U.S. counterparts with fewer overseas opportunities will not feel as strongly.

The eventual technological "balance" in a liberal world telecommunications market is impossible to predict. But in the interim, the industry that must deal with the most direct competitive challenge should have the strongest incentive to innovate, and from among the firms in the group that succeed should emerge many of the world's technological leaders. As other countries then open their markets to a broader range of participants, these "leaders" should find themselves well-prepared to face the rigors of new entry.

In sum, the immediate technological consequences of uneven (across the major supplier countries) liberalization may seem ambiguous, with overall gains being made but with the relative position of an open market producer proving more difficult to sustain. The longer-term results, however, are more clear. The experience of competition should place successful manufacturers in a strong position to cultivate the new opportunities that materialize as liberalization unfolds abroad. Ultimately, should a genuine global telecommunications market emerge, these firms should figure prominently among those who provide the technological and commercial leadership of the future.

Other Peculiarities of the Telecommications Equipment Market

International competition in telecommunications equipment is also colored by other unique features of telecommunications demand and trade:

1. Major projects make up a substantial proportion of telecommunications equipment demand (especially in the developing world, where first-time systems are common).

2. The direct statistical implication of this market feature is that trade patterns (particularly on the bilateral and regional levels) often prove highly irregular. As an example, U.S. telecom exports to South Korea quadrupled between 1980 and 1982; however, they are

expected to return soon to improved (vis à vis the 1970s), but less dramatic, levels.

3. The main competitive implications are also predictable:

a. services and follow-up provisions figure prominently in customers' choice of suppliers (see "the Systems Solution" below);

b. bidders on a project will often set an initial goal of a "foot-in-the-door", absorbing less attractive short-term financial conditions for the sake of more profitable, long-term follow-up;

c. the "conditions" of such large-scale bids (export licenses, export financing provisions, etc.) critically influence final purchase decisions, and their importance to a supplier nation's equipment industry can lead to an international competition between public support organizations (over and above the firms involved).

4. Also reflected in the structure of international competition is the fact that, while clearly a high-technology industry, telecommunications equipment embodies a wide range of technical capabilities and sophistication in its many products. This can lead to significant variations in supplier countries' standing from one market segment to another (cable vs. satellites, for example).

As noted, these characteristics complicate any analysis of telecom trade. But they also must play an essential background role, in public choices concerning government's place in the industry and in corporate choices on questions ranging from research emphasis to marketing strategy.

The Systems Concept

As if in anticipation of trends that now epitomize other high-technology sectors (such as computers and robotics), telecommunications equipment has long been a systems-oriented industry. The original justification for a legalized service monopoly was the need for a telecommunications network—read system—that functioned as an efficient unit. While recent liberalization has redefined the limits of its applicability, the concept lives on in a simple, enduring fact of the marketplace: customers still buy to meet a need or solve a problem, and in telecommunications, that means buying a system. In response, young fast-growing service companies have pursued a measure of vertical integration in order to join equipment and support into a more marketable package. Customer requirements have compelled these firms to develop some internal equipment capabilities and sources, and those who emerge with well-tailored systems offerings could enjoy a strong position in some major segments of both the equipment and services markets.

Just as telecommunications systems have (in accordance with age-old reasoning) fused equipment and service into single packages, so has technological improvement served to blend the field as a whole with related computer products. Data communication and data processing provide the most obvious example of how these once distinct sectors overlap and interlock. Even where conceptual distinctions may remain clear (data transmission vs. data manipulation), the practical aspects of carrying out apparently discrete functions still weaves them together into inextricable processes, technologies, and systems. In Japan, most of the major telecommunications companies are also leaders in the computer industry—Fujitsu, Hitachi, NEC, Oki. There, as in the U.S. and Europe, the successful joining of those two fields will prove a key factor in determining the international competitiveness of a nation's telecommunications equipment manufacturers.

Skills

Considerable attention has recently been devoted to a looming shortage of the skills required by the telecommunications industry and related high-technology sectors. Concern has centered around:

1. the declining number of students graduating with an emphasis on engineering, the sciences, and mathematics;

2. the dwindling population of qualified teachers and professors in these critical areas (due largely to the salary gap between academics and private industry); and,

3. the gradual deterioration of available instruction in these quantitative fields (for reasons of both inadequate staff and aging facilities).

Well-developed human resources have been an important key to American pre-eminence in "knowledge-intensive" industries like telecommunictions equipment and services. The indispensability of a well-trained pool of eventual contributors has led countries as diverse as Singapore, West Germany, and Japan to devote considerable effort to the cultivation of such capabilities. Each is acutely aware that a shortage of skilled technicians can impose limitations on high-technology development no less serious than financial or production constraints. Several U.S. corporations have already, in recognition of this problem, provided assistance to various universities for purposes ranging from overall technical education to equipment moderanization; in some instances, they have even established their own institutes for instruction purposes. The options section discusses certain additional possibilities for USG action on the training and educational issue.

Technological Change

By definition, a high-tech industry like telecommunications involves steady, and sometimes spectacular, technological progress. Major research organizations like Bell Laboratories have earned impressive reputations as key contributors to this advance; smaller firms and creative individuals have also furthered their innovative traditions in ways that permanently transformed the industry.

Today, both for telecommunications and for high-tech in general, two trends predominate. First, new concepts and products are appearing in the marketplace with greater frequency than ever before. But at the same time, the more stringent requirements for success in this dynamic competitive environment have accelerated the process of technological diffusion, shortening the effective commercial life of any particular breakthrough. On the one hand, technological eclipse becomes an immediate concern under such rapidly changing conditions for those who, however briefly, might fall behind. On the other hand, a technological edge means less and less to those who might consider themselves leaders. Obviously, an intensive research and development effort is required of those who would become and remain prominent suppliers of telecommunications

193

equipment. In certain areas, such as fiber optics and satellite communications, R&D results could prove decisive for a particular firm. But the other ironic implication of these trends for less exotic areas of high-technology competition is that the non-technological aspects of commercial success—service and support, marketing, product reliability, pricing strategies, financial strength—could just as often determine exactly who the leaders will be.

Research and Development

As always, research and development activity remains a critical ingredient of an internationally competitive telecommunications industry. The "technological acceleration" that has come to characterize most sophisticated sectors ensures that only the innovative survive.

Overseas, research and development in the telecommunications field has drawn heavily upon public support, largely because of extensive government involvement in the operation and regulation of national telecom markets. In relative terms, the U.S. private sector has assumed a greater share of the industry's R&D responsibilities, a pattern reinforced by recent liberalization measures, budgetary trends, and tax policy changes. Over and above AT&T divestiture, the main forces at work in the area stem from the Economic Recovery Tax Act (1981), which, in conjunction with condensed government research activity (for FYs 1981-3), sought to encourage telecommunications firms to increase further their R&D efforts. The three basic measures included in ERTA for this purpose were:

    1. tax credits of 25% of any increase in corporate R&D expenditures (through 1983);

    2. a two-year suspension of allocation rules governing tax treatment of research and development outlays; and,

    3. accelerated depreciation for R&D facilities and equipment.

Specific data on the impact of these provisions is as yet unavailable, but despite some conflicting early reports, it is hoped they will catalyze an increase in research and development activity.

At the same time, broader interpretation of antitrust statutes may now enable the formation of selected private-sector R&D consortia, along the lines of the Microelectronic and Computer Technology Corporation (MCC).[10] Such groups will attempt to exploit the economies of scale and risk-reduction opportunities that collective efforts in basic R&D may provide. Should this effort in fact establish a trend, it could mean more frequent inclusion of smaller U.S. companies in the long-term research activities essential to the continued growth and competitiveness of the U.S. telecommunications equipment industry.

Foreign Targeting Practices

Because of extensive government involvement in telecommunications activities (an involvement which historically has not emanated from an exclusive interest in high-technology promotion), application of the term "targeting" must be based upon careful and judicious review of individual markets and their evolution. But despite the occasional awkwardness of discussing industrial policy in a field so recently regulated by governments worldwide, the lure of high-technology development appears, in several instances, to have provided new incentives for maintaining otherwise unwarranted control over a national telecommunications sector. Some indication as to the extent and legitimacy of this control is provided in other sections of this report (cf. the regional market and world trade discussions). The thorough, country-by-country analysis that specific policy judgments require would entail a far more extensive treatment than this context allows.

Nevertheless, certain general concerns can legitimately be raised. The transition from market regulation to market protection is an easy one. The standards systems and quasi-public monopolies once established to facilitate the rise of a functional telecommunications network can also be applied to impede access and inhibit competition. The telecommunications market, in both equipment and ser-

10. The MCC venture will apparently be constrained as follows: a) it cannot be a profit making enterprise; b) no firm will be allowed more than a 10% interest; c) the Justice Department will also monitor the overall corporate membership, the identity of companies participating in particular projects, and whether or not the risks involved in MCC's efforts are sufficient to justify joint efforts.

vices, has itself changed dramatically in recent years, above all due to changing technologies, applications, and demand. But predictably, public policy treatment of the industry has evolved at a slower rate, with considerable variation in the adaptability of major countries to the new conditions and opportunities that market liberalization would imply.

Amidst the motives for the reluctance of some to change is targeting, disguised by its peculiarly passive form in an industry from which governments are generally withdrawing. This can take the form of special tax treatment, joint (public/private) R&D, direct funding through loans/grants, etc. But for the moment, these types of distortions present less direct and less onerous interference than the traditional barriers of restrictive standards and procurement. Though defensive in nature, these targeting practices account for much of the asymmetry and trade imbalance that characterize the current world market, and they constitute a most serious issue for American manufacturers of telecommunications equipment. Unless U.S. producers obtain equitable access to foreign markets, the full benefits of liberalization may not be realized. The most pressing challenge before USG officials is to secure such access. Then the more ambiguous effects on competitiveness in telecommunictions of more typical high-tech support programs can be analyzed and addressed.

Options

A Discussion of the Pros and Cons of Proposals for USG Action as Recommended by a Variety of Sources.

Research and Development

A primary issue of great interest is the level of R&D activity in the United States. Several recent policy changes have acknowledged its importance for the nation's future, especially in high-technology areas, but some concern remains over the long-term implications of inadequate R&D expenditures.

The telecommunications industry has long been a leader in terms of its R&D outlays. Revisions in the tax treatment of R&D (embodied in the 1981 Economic Recovery Tax Act) were designed to further stimulate such expenditures. The accelerated depreciation schedules included for R&D equipment should provide significant and secure incentives for corporate investment in this area. The accompanying R&D tax credits, however, may need elaboration to ensure their effectiveness.[11] Insofar as firms' decisions on R&D allocations require long-term planning and more extended lead-times, two years (the applicable period of these current measures) may prove inadequate for generating a broad positive response. A longer-term provision of this type may prove desirable. A second possible shortcoming of the stepwise R&D credit may be its lack of stimulus for the young, fast growing companies that so heavily populate research-intensive sectors, and from which an impressive proportion of technological innovation has emanated. The simple incremental approach embodied in existing legislation may provide the least benefit and incentive to many of those active in the area of policy concern. Revisions that structure into the formula credits for a baseline, dollar-amount R&D increase (on top of which the 25% schedule would take effect) might somewhat alleviate this problem.

Another set of policy developments in the R&D field involves more open interpretation of anti-trust regulations (see page 195 for discussion). Recognizing that some legitimate economies of scale can be realized through limited inter-firm collaboration in areas of basic research, the Justice Department has given qualified approval to the establishment of the joint venture MCC, the Microelectronic and Computer Technology Corporation (again, see p. 195). Many observers feel that this could represent an important first step towards similar cooperative activity in other high-technology sectors. But it appears that before any field can reap the full rewards of this new understanding, the ground rules will need to be clarified and se-

11. Included are both the "25% of increase" credit and the reallocation of international credits assigned under Code Item 861.

cured. The lack of detailed and defensible preconditions will likely deter many valid participants from joining a collective undertaking of this type. At this point, considerable discretionary/interpretive power remains with the Justice Department, the courts' position on such ventures has yet to be clarified, and no protection from civil suits has been provided. In the face of such impediments, legislative action may emerge as the only mechanism able to catalyze full use of this collaborative opportunity.

Personnel

A second, long-term problem for both the telecom industry and the economy as a whole is the "skills shortage" (see page 192 for a discussion of this issue). Increases in the number of new scientists, mathematicians, and engineers have not kept pace with a growing field's demand for this type of trained personnel. This has in turn led to a depletion of the ranks of qualified instructors remaining in academics. And to complete the cycle, this shrinking number of teachers (in both secondary schools and universities) is less able than ever to educate the larger numbers of trained students needed by high-technology sectors.

One solution to this problem is to provide a greater network of government support for education and training in the areas of concern. Specific recommendations are to:

A. increase public funding for discretionary improvement and enlargement by educational institutions of programs aimed at the training of scientists, mathematicians, and engineers; and/or

B. lend more directed public assistance, in such areas as teacher salaries, for the purpose of encouraging qualified academics to remain in their positions as instructors; and/or

C. provide incentives to the private sector for further corporate training of the necessary personnel; and/or

D. provide incentives for broader and indirect private sector assistance to educational programs in the maths and sciences (such as the Computer Equipment Contribution Act considered by the Congress in 1982).

These kinds of measures would both facilitate scientific education

198

and help pique the interest of greater numbers of prospective students in the designated fields. It could also assist in retraining of workers whose skills have become obsolete because of shifts in the U.S. production base.

However, special caution would need to be exercised to avoid intensifying the current competition between industry and academia for skilled people (see, for example, the proposed salary assistance for teachers). Also, increased budget support would be required under any of these programs unless current resources were transferred from the liberal arts to the sciences, a move that could generate considerable opposition from other affected interest groups.

Another solution is to leave the necessary adjustments to the marketplace. For some, this may offer a more efficient alternative to the kinds of government involvement implied in the policy options outlined above.

But, the adjustments needed to restore equilibrium between the supply and demand of technical skills under a laissez-faire approach may require an inordinate period to complete. The shortage of qualified personnel is an immediate problem which, if not addressed soon, could have adverse long-term consequences for the American economy. In other words, the employment market may function inefficiently in translating sudden changes in demand through educational institutions into shifts in the training and eventual supply of properly equipped graduates.

Countering Foreign Competition

A third area of concern to the telecommunications industry is the dual problem of foreign targeting and the restriction of market access. Considered together here because they so often coincide, these practices could seriously affect the competitive position of U.S. equipment manufacturers.

Adopting comparable practices could be formulated to spur development in any of several areas of the telecommunications equipment industry, presumably placing U.S. firms on an "equal footing" with their foreign counterparts. Comparable targeting practices could conceivably encourage risk-taking among producers and avoid undesirable waste and duplication of R&D efforts. Outright protection could perhaps result in an eventual dismantling of selected foreign industrial policy programs, if this step is accepted as the price for regaining access to the U.S. market.

Certain minor types of support (especially basic research in government and university laboratories) and special tax credits for R&D are already in place. More liberal interpretation of anti-trust regulations has also enabled some joint research efforts to be organized among high-tech firms. More sweeping measures would require a fundamental reversal—in the current philosophy of U.S. business-government relations in general, and of telecom competition in particular. Such changes could also either a) shift competition in the equipment field from production programs to support programs as other countries in turn attempt to provide the most generous terms for development or b) intensify the use of less palatable trade barriers, such as tariffs, quotas, etc. Again, industry-specific USG policies could invoke demands for equal treatment across a whole range of American sectors that also feel threatened by foreign support.

Exercise of the market protection option (including such practices as closed national procurement) could:

A. mean a sacrifice of the broader gains from liberalization (primarily in terms of consumer welfare and the overall efficiency of capital distribution);

B. have an ambiguous effect upon the U.S. firms it was designed to strengthen (since their long-term competitiveness may be depleted without the challenge of foreign firms, since many are active in international joint ventures within the U.S. that depend upon open trading channels for their success, since others are multinational firms who would be prime targets for any retaliatory moves on the part of foreign governments, and since this would surely paralyze any global shift towards greater openness in telecommunications, a shift that could relatively favor the technologically superior U.S. industry).

C. would fail to address the question of third-markets—and protective diversion of foreign exports could erode U.S. market share abroad; and

D. raises the specter of a full trade war, an eventuality that could seriously damage the U.S. economy across a much broader range of products and industries.

A vigorous U.S. program to *counter targeting programs* through strict enforcement of U.S. trade laws (Section 301) could produce case-by-case agreements as to how an equitable trading environment could be restored. Though few precedents involve the specific background questions that might surround a telecom case, this has historically been a successful process, with only rare invocation of Executive Authority to impose unilaterally reciprocal restrictions. Above all, active enforcement would lend integrity to the legal structure now in place.

Most U.S. trade laws only emphasize temporary relief and adjustment assistance where damage is found, offering little to actually discourage targeting practices. The resources required of firms to pursue trade action cases, and the often lengthy period between violation and judgment, may discourage many (particularly smaller companies) from invoking what provisions are available. While several complaints are still in decision at this time, Section 301 (the mechanism relevant for most targeting problems) would also appear not to deal with the question of third markets, and many nations may well decide that the benefits of promotion policies still outweigh the costs of American enforcement. Finally, there remains some question as to the GATT-legality of certain responses of this type.

Negotiation through bilateral and multilateral channels for country by country removal of the most restrictive practices would hopefully give rise to a consistent and principled U.S. strategy for dealing with these types of restrictions in an international context. It could introduce an essential measure of equity and efficiency into the world telecommunications market by restoring a fully competitive atmosphere for international trade and investment.

The U.S. approach to discussions of this kind would have to take into full consideration the range of other economic, political, and security interests at stake. Because these may vary greatly from case to case, it could prove difficult to develop any set of policy positions on restrictions and targeting that appears coherent and non-arbitrary.

Export Controls

A final concern for industry and government alike has been the effect of export controls (whether COCOM or unilateral restictions) upon telecommunications equipment firms and sales.

USG policies in this area will clearly be attempts to balance sometimes conflicting objectives, and any effort to dictate a binding solution in this context would prove highly problematic. However, the upcoming renewal of the Export Administration Act does provide an opportunity for debate over the methods and content of export controls. As part of the review, it could be appropriate to:

A. re-emphasize the priority of technology transfers over product transfers as a guiding principle for security concerns.

B. underscore the broad damage to commercial interests that results from perpetuating the United States' reputation as an "unreliable supplier".

C. highlight the fact that export markets in high-technology fields also contribute to national security by expanding the U.S. military/industrial base.

D. urge all policy-makers involved to give full attention to the competitive interests of the relevant manufacturers, noting in particular the disproportionate burden that export restrictions can place on smaller telecommunications firms.

E. formalize this advocacy role by involving Commerce Department industry and trade policy specialists in future discussions of telecommunications trade controls.

Telecommunications Industry Overview

BY JOHN SODOLSKI
VICE PRESIDENT,
TELECOMMUNICATIONS GROUP
ELECTRONIC INDUSTRIES ASSOCIATION

Our definition of the telecommunications industry includes the following equipment and systems: switching; transmission, cable and wire; outside plant; PBX, key systems and telephones; data communications equipment; satellite communications equipment; mobile radio equipment and systems; broadcast equipment; record terminals and other premise equipment.

The market for telecommunications equipment in the United States is generalized at something over $23 billion in 1982. The telecommunications market for all of Europe in 1980 was in the neighborhood of $10 billion. The telecommunications market in Japan as represented primarily by the Nippon Telegraph and Telephone Co. (NTT), and some other small markets, was a bit over $4

203

billion in 1981, and not the entire $4 billion is telecommunications. The non-U.S. world market for telecommunications equipment and systems is estimated at just under $40 billion for 1982. Estimates for 1987 indicate a U.S. market of about $34 billion and a world market of just under $60 billion.

National markets for telecommunications equipment and systems in the industrial nations of the world are largely closed to U.S. manufacturers and suppliers. With the exception of the U.S. and Canada, every major industrial nation's telecommunications system is controlled and operated by a government entity, usually a post and telecommunications authority, PTT.

Each nation's voice and data networks are operated by the PTT, and only to a limited extent is certain ancillary or "interconnect" equipment purchased and utilized by customers of the PTT, rather than the PTT itself.

In addition to the control and operation of the network by the various PTTs, each telecommunications authority usually has one or several major domestic suppliers from whom it purchases the preponderance of its requirements. Often, if not usually, R&D toward new telecommunications equipment, systems and usage is funded by the PTT, and that development work is handled by the various chosen instruments.

— This R&D work usually leads to production contracts, often long-term contracts lasting for as much as five or seven years. It is almost unheard of for a production contract, much less an R&D contract, to be awarded by a PTT of an industrial nation to a foreign supplier. In a very few instances, certain European countries have purchased American equipment where they felt a real need for the U.S. high-tech or where it seemed politically expedient to do so.

In some countries, especially Japan, this has never been the case. They have done all development and all procurement with domestic firms.

This was to have changed with the signing, in December of 1980, of the Nippon Telegraph and Telephone Public Company (NTT) agreement between the U.S. and Japanese Governments. The agreement was the result of years of negotiations. The agreement is cumbersome and complex, and at this point has yielded no significant

204

sales from the U.S. to Japan. At the same time (1980-1981) Japanese exports of communications equipment to the U.S. had increased by over 60 percent, from about $.4 billion to about $.66 billion. We estimate telecommunications imports from Japan will be more than $.8 billion in 1982. U.S. telecommunications exports to Japan, in the same categories, for the same years, have been in the neighborhood of $40 million.

In addition to the institutionalized barriers to sales in developed and industrial nations, there is a litany of problems which encumbers even the minor amount of telecommunications trade which might obtain against the background of PTT controls. These impediments include a number of non-tariff trade barriers, not the least of which is type-acceptance or type-approval of telecommunications equipment before it may be used within that nation.

The entity which approves equipment in almost all cases is the PTT, who quite often will protract the process so as to discourage U.S. and other firms from even attempting to obtain such approval. This process is especially effective in protecting the ostensibly open "interconnect" market mentioned above.

Some U.S. firms have subsidiaries in Europe and operate within those markets with varying degrees of success through those subsidiaries. It is unusual, however, that equipment and systems are sold to the European market directly from the United States.

The net result of this is that Western European markets are de facto open to U.S. commercial telecommunications sales. The Japanese market at this point appears closed for any non-Japanese equipment and systems. The December 1980 NTT agreement may change that, but at this point—two years after the agreement was signed, after several years of negotiations—there still have been no significant sales.

This must be contrasted with the U.S. market for telecommunications equipment and systems, which is almost completely open to all foreign competition. With a secure domestic market base untroubled by foreign competition, our international industrialized trading partners are able to provide and sell high-tech equipment in the United States and to price flexibly while we are unable to do the same in their markets.

TELECOMMUNICATIONS

The imbalance and disadvantage to U.S. communications manufacturers is obvious. An equally obvious and seemingly appropriate answer would be equivalence in competitive access to telecommunications markets of our industrialized trading partners. However, even equivalence has its problems and imbalances.

If there were full and equitable telecommunications trading arrangements, the United States has a far larger, more homogeneous and better organized market than any market available abroad to U.S. manufacturers, and those markets are not at this point available.

The U.S. posture internationally in telecommunications trade, as in high-tech generally, is a growing problem. I have seen the attitudes of executives in the telecommunications industry change from concern just a few years ago, to dismay more recently, and many are now verging on anger at the trade situation.

Many of us believe telecommunications to be the premier high-tech industry of the balance of this century. Some have suggested telecommunications bears the same relationship to the next twenty years that steel had for the last two decades of the last century—the stuff that nations are made of. If that is the case, it's vital that the U.S. not allow its technological and production base to be eroded by unfair trade practices. Several courses of action have been proposed in the U.S., ranging from the usual trade remedies, to the serious consideration of so-called reciprocity legislation in the Congress.

We must stop tiptoeing around this problem, and recognize the nature and seriousness of the problems facing the telecommunications industry. Our industry is willing to meet competitiors anywhere, anytime, and to win fair competition based on price, quality, reliability, or any other objective criteria. However, to do that, we must be able to meet our competitors on their ground as well as our marketplace, and the rules should be understood, and be the same for all.

The Need to Open Closed Markets

BY MR. RICHARD MOLEY
VICE PRESIDENT, MARKETING
ROLM CORPORATION

ROLM Corporation is a U.S. company, headquartered in California's Silicon Valley, with annual sales running in excess of $400 million. ROLM entered the telecommunications market in 1975 with a computer-controlled, fully-digital PBX and since then has become, we believe, the largest manufacturer of digital PBX systems in the U.S. We have expanded our sales activities in the international market and have had success selling in open and relatively open markets, e.g., Hong Kong, Mideast, Australia. In March 1983, we received approval to sell our systems in Japan, a previously closed market, and have had some success there too. We are very concerned that many of the large markets are effectively closed and that protectionist sentiments are increasing in some markets that are currently open.

TELECOMMUNICATIONS

In discussing the subject of closed markets, it is important to segment the huge telecommunications market because the issues involved in each segment can be quite different. Although there are many ways to do this, a useful one for the purposes of our discussion is to consider the following segments:

1. The provision of voice and/or data transmission services.

2. The manufacture and supply of equipment used primarily in the provision of transmission service, e.g., central office equipment, transmission equipment, cable, etc.

3. The manufacture and supply of terminal equipment for business and consumer use which is often referred to as customer premise equipment (C.P.E.).

The size of these segments is generally in the order listed with transmission services being the largest and CPE the smallest.

In most large developed countries transmission services are provided by governmental bodies (PTT's). Central office and transmission equipment is manufactured by a small group of domestic companies (or subsidiaries of approved multinational companies) to the specifications of and upon request by the PTT's. C.P.E. equipment is manufactured to the specifications of the PTT's and sold either through the PTT or directly to end users on an interconnect basis.

In the U.S., transmission services used to be provided on a monopoly and regulated basis by ATT and independent telephone companies and were as effectively closed as that of any other country. Although a combination of FCC decisions and the pending divestiture of the BOC's has opened the long distance portion of this segment to competition, it is unlikely that non-U.S. companies will become a fact here. Nobody has any realistic expectations that other countries will open this segment to international competition in the foreseeable future.

The second segment, while theoretically open, was substantially closed to both domestic and foreign competition due to the fact that both ATT and GTE bought most of this type of equipment from their wholly-owned subsidiaries: Western Electric and Automatic Electric, respectively. The pending divestiture of the BOC's offers the possibility that a substantial part of this market will be open in the future to other domestic and foreign competition. It remains to

208

be seen, however, whether the BOC's will change their historic dependence on Western Electric and, of course, neither ATT nor GTE is under any obligation to purchase from other than their own subsidiaries.

The third segment is open to both domestic and foreign competition as a result of the original 1968 carterphone and subsequent decisions of the FCC. In fact, it was the opening of this market that stimulated ROLM to develop our CBX system. Today both Canadian (NTI & MITEL) and Japanese (NEC, FUJITSU, OKI, IWATSU, etc.) suppliers have significant sales in this segment.

It is vitally important for the health of the U.S. telecommunications manufacturing industry that the international markets be opened for the following reasons:

1. The international markets are very large and have the potential for large sales and exports for U.S. companies.

2. The U.S. equipment markets are open in various degrees. If we cannot sell in the domestic markets of our international competitors, we run the risk of having these competitors use their protected markets to provide product development funds and profits. They can then treat the U.S. market as an incremental one with incremental pricing which we cannot meet. I am optimistic that we can compete aggressively. We have the finest telecommunications technology, but we need assistance from the U.S. Government to open the doors.

We face four different types of barriers:

1. political, where a government decides who produces what;

2. economic, where there are high tariffs and/or local content requirements which create a very high risk for the U.S. manufacturer;

3. technical, where either specifications are unavailable to foreign suppliers or the cost of meeting unique specifications cannot be justified due to market size and potential;

4. technology transfer barriers generally seen in newly industrialized countries.

As ROLM does not manufacture central offices or transmission equipment, but only business customer premise equipment, I can only speak from experience in the third segment defined above.

TELECOMMUNICATIONS

There is considerable common ground between the second and third segments which we might discuss in the discussion period which follows. The recent U.S./Japan trade agreement with Nippon Telegraph and Telephone Corporation is an excellent example of how a trade arrangement has served to help ROLM and potentially other U.S. companies to penetrate the Japanese market—the second largest telecommunications market in the world.

While Japan has constructed a myriad of trade barriers, it is, as far as we know, the only foreign country that has agreed to open its telecommunications market to foreign sales under the procedures of the Government Procurement Code. It provides for improved access to Japan's interconnect market which is regulated by NTT.

ROLM's initial contacts with NTT occurred in the summer of 1979 while trade negotiations were in progress. We quickly found that while there were no serious tariff barriers and there were technical specifications for an analog private branch exchange, there were no specs for digital PBXs. Furthermore, there were no procedures to guarantee any NTT evaluation of products from any non-Japanese supplier who applied for approval to sell—a classic technical/political barrier. NTT planned to publish specifications for digital PBXs to connect to their network and had requested the four major Japanese PBX manufacturers (part of the NTT "family") to make proposals. The result was four significantly different approaches resulting in NTT delaying introduction of digital technology for business communications. As early as 1979, we met with NTT in order to understand the existing policies and procedures, the format of the approval process in general and to meet key members of NTT's technical staff.

Immediately after multilateral trade negotiations were completed, we discussed with NTT the digital PBX specifications and suggested some changes based upon the highly regarded Australian specifications and procedures. Trade negotiations had wisely called for formalized type approval and procedures, which were eventually established. NTT went out of its way to make it known that procedures were being put into place and that complete digital specifications would be published. They sent a team to tour the U.S., as part of the agreement signed in December 1980.

TELECOMMUNICATIONS

Since we had started work on the project well over a year earlier, we were now able to participate in reviewing the specs for the digital PBX, and in the spirit of the agreement, NTT accepted the concept of a "black box" approach rather than a specific internal design approach. In late summer, 1981, NTT published the digital PBX specifications in Japanese and an English version in September. By this time ROLM had its hardware and software design nearly completed, even though several modifications were required to meet the then published, final technical parameters. In December 1981, we submitted the required documents for NTT type approval (measurements, computer printouts, block diagrams, diagnostic manuals, etc.). NTT was required by the formal agreement to accept or deny documents submitted within 60 days. With more than 3,000 pages of translated documents approved, we submitted our product for physical examination in late February, 1982. Official type approval for all ROLM private branch exchanges by NTT was formally announced on March 26. NTT's requirements were tough, but fair, and in line with what we expected from a sophisticated telephone company. The commercial section of the U.S. Embassy in Tokyo was helpful throughout the process—both through their knowledge of the Japanese culture and their basic understanding of the technology involved. Our consistent suggestion, for instance, that the digital specifications for a PBX be an external one was carried to NTT on a persistent basis by personnel from the commercial section.

We appointed Sumitomo Corporation, one of the largest companies in Japan, exclusive distributor. Sumitomo, in turn, appointed Omron Tateisi Electronics Company, a recognized leader in the electronics industry, to spearhead the sales and service of ROLM products. Both companies helped tremendously in the efforts to obtain type approval and to identify final specifications within the short final three-month time frame. As eighty-five percent of Japan's PBX sales are made by interconnect companies, we are primarily pursuing the interconnect business, although we have responded to the NTT EP10 tender. While all the preliminary pieces are in place in Japan today, we must now sell to the end-user—businesses in Japan. The task is not finished; it's really in its infancy. Japanese customers are highly nationalistic and we face fierce competition from the domestic supplier, who of course, would prefer that we not succeed. We have had some important successes, most notably a large contract with JAL and are cautiously optimistic that we will succeed.

Why do we like the agreement negotiated with NTT?:

■ It is a type approval agreement which assures the supplier that he has free access to the market once his product has been approved and as long as he does not make major modifications.

■ The approval authorities must approve or disapprove a submission within a short time frame.

■ Disapproval must be documented with a time allowed for correction and without the necessity of having to reapply.

■ Physical examination and testing of the equipment is handled in the same manner.

■ More important, testing results obtained in the U.S. or elsewhere if obtained by an "accredited" agency or done in accordance with methods accepted by the international telecommunications union in Geneva, Switzerland, are submittable and valid.

■ The agreement is short, precise, and clear.

What can we all learn from these experiences in Japan?:

1. The U.S. Government must continue to find means, through negotiations with other countries, to discourage protectionist measures and Government-backed export incentives which we have seen, and continue to see, in Europe, the newly industrialized countries, and even our neighbors to the North—Canada.

2. The NTT/Government Procurement Agreement is an important test case. The U.S. Government must take similar steps for such agreements in other countries in a more aggressive manner. There must be negotiations for as realistic and helpful agreements as was accomplished in Japan.

3. When specifications are being developed, emphasis should not be on internal design, but rather the "black box" concept; that is, what goes in and comes out, not what's included inside—external parameters, rather than internal guts. This would take away almost all the barriers I had referred to as technical barriers earlier.

4. A published procedure for type approval is essential. Obtain-

ing approval on each system would be so costly as to discourage attempt.

5. A knowledgeable, interested and aggressive commercial section in U.S. Embassies is extremely helpful.

While the GATT Agreement is designed to lower international barriers to trade, we do not see many inroads in the European Communities or the newly industrialized countries. One gets the impression most countries are constantly looking for new ways of closing their markets, especially during this recessionary period. The continued health and industrial competitiveness of the U.S. telecommunications industry is threatened by the existence of closed markets. To add teeth to the Commerce Department efforts to negotiate agreements to eliminate foreign restrictions, the U.S. may, as a last resort, have to adopt the concept of reciprocal treatment. While reciprocity could reverse the progress toward open trade, we believe adopting some of the concepts of the Danforth Bill will indicate the U.S. is serious and, in the spirit of the GATT Agreements, wants fair and free trade, worldwide.

Targeted Industry Impact on the Domestic Telecommunications Industry

BY JOHN F. MITCHELL
PRESIDENT
MOTOROLA, INC.

Throughout most of the developed world, the telephone has evolved in a highly protected environment. The government operates the telephone system in most countries through their PTT and utilizes a minimum number of closely tied equipment suppliers. These suppliers have developed significant exports to other less developed countries that do not have a local telecommunications manufacturing industry. In the U.S. the telephone industry is changing very rapidly as a result of recent FCC action and the government's divestiture agreement with AT&T. The U.K. has a similar program, but most other developed countries, including Japan, remain nearly a closed market.

Private two-way radio services outside of the U.S. have been severely limited by radio equipment and frequency regulations. In the U.S. private radio services have been extensively developed because of a comparatively more liberal policy for equipment and frequency allocations. In recent years new types of service such as cellular radio and SMRS, or shared systems, have opened the import market to a much greater degree by concentrating the purchasing power for radio equipment. All of these changes play into the hands of offshore industry already practicing the essence of targeting. That is: protected home market, government subsidized procurement and R&D, export incentives, and government sponsored collaboration. There is a tremendous dichotomy between the open, rapidly deregulating telecommunications market in the U.S. and the continuing national policy of a closed protected market in other countries, especially Japan.

Many governments have viewed their telephone industry and now radio as a natural vehicle to develop a strong local semiconductor and industrial electronic capability. Integrated circuits is another area where government support in many countries adds distortion to the world market for telecommunications. The Japanese government program, targeting the semiconductor industry, has been the most successful of all these efforts.

My colleagues have described the protection and support that foreign governments extend to their domestic communications equipment producers. International competition is probably more distorted in telecommunications than in any other industrial sector. U.S. firms must now compete without the benefit of the massive subsidies foreign competitors receive through discriminatory procurement by PTT's, government R&D support, and plentiful official export credit support. The net effect of these subsidies is to substantially increase the risks for U.S. producers in this industry for non-market reasons. Every U.S. producer recognizes and accepts the inherent risks of competing in an open market, but we do not believe we should have to accept the additional risks that foreign governments impose on us through their targeting policies.

The denial of access to our foreign competitors' home markets is a major inequity, as is the pattern of government supported export to Third Country markets. These inequities are being and should continue to be addressed through U.S. government efforts to negotiate greater access to our competitors' home markets and to limit distortive competition for Third Country markets. But even if those efforts begin to succeed, their positive effects will not be evident for some time, as the NTT agreement graphically demonstrates. What is vitally and immediately important to the future of the U.S. communications equipment industry is what the U.S. government does about unfair competition in the U.S. market.

The tremendous dichotomy between the openness and rapid deregulation of the U.S. telecommunications market and the virtually complete closure of our competitors' home markets is no longer tolerable. The U.S. market will be vulnerable to an unprecedented degree during the coming years. U.S. producers face a situation in which their major competitors will still have a de facto secure home market while ours is under severe attack.

We can readily expect that the Japanese will be the leaders of their assault. They have demonstrated a capacity and willingness to dominate one industry after another through government support and protection. Telecommunications is no execption. NTT has, through a series of five-year plans over the last thirty years, given Japanese firms a strong technological and production base. MITI has also been active in developing certain technologies like VLSI and fiber optics that are vital to modern communications systems. The series of temporary measures to "rationalize machine tools and industrial electronics" administered by MITI over the last twelve years is another important role in building their telecommunications industry.

The result is an imbalance in U.S.-Japanese trade that is increasing very rapidly. A favorable Japanese balance of 20 to 1 simply cannot be explained by any reasonable examination of the relative competitiveness of the U.S. and Japanese industries. We have proved that in paging. If the U.S. government does not act quickly and effectively, we are convinced this imbalance will grow at an accelerated rate during the coming years, with disastrous consequences for U.S. producers and U.S. technological capabilities.

Motorola is doing everything it can to overcome this fundamental imbalance in this competitive situation. We are doing many things within the company to improve our competitiveness. We have made major commitments to serve the Japanese market and have achieved some modest success in selling pagers to NTT. We are also well along in qualifying to sell NTT car telephones. We have sold semiconductors and two-way radios in Japan for over ten years. We manufacture semiconductors in Japan, and we know the problem of market entry in Japan.

Nothing so graphically illustrates the inequity of the situation as our experiences with pagers. After three years of intensive effort, we succeeded in selling pagers to NTT. NTT changed their usual award procedures so that each bidder receives one-sixth of the annual volume. We could not underbid our competitors in Japan or offer a unique product or feature and get a larger volume.

Shortly after receiving our initial orders from NTT, a Japanese supplier of NTT pagers entered the U.S. pager market. The Commerce department has recently determined that firm was selling at less than fair value by an 88 percent margin. We, of course, had to drastically cut our prices to meet this competitive threat resulting in very large losses in revenues for Motorola. In fact, within a couple of months we lost more in the U.S. market than we can expect from our years of effort in the Japanese market.

This is not the first time that Japanese electronics firms have dumped into the U.S. market in an effort to dominate a market. In the case of color TV, a business which we were in at the time, these tactics, by some of the same companies now targeting telecommunications, were very successful, particularly when the U.S. government failed to completely enforce the antidumping statutes. To my knowledge, the $400 million dumping duties still have not been paid.

While I have talked primarily about Japan, the situation is similar with respect to our European competitors although they may not constitute as grave a threat in the U.S. market.

What we face is a pattern of competition where certain major foreign markets are effectively denied to us for the foreseeable future, where competition for Third World markets will be intensive and involve foreign government financial and political support, and our competitors will have free access to the U.S. market limited only by our ability to chase them down in product after product under the unfair trade statutes. This is an extremely unfavorable environment for U.S. firms to maintain their competitiveness and technological strength.

For these reasons the U.S. communications equipment industry feels unprecedented action is warranted. In 1982, the EIA Telecommunications Group supported the strong reciprocity provision embodied in telecommunications legislation (S.898). That provision would deny access to the U.S. market to countries that deny access for U.S. products. This would take great judgment to enforce since there is a monumental difference between the approval process at the FCC for entry into the U.S. market and the process of approval, for example, of one two-way radio into Japan.

We understand the risks of this approach to the broader trading system, but we think the situation in communications equipment is so serious that special actions are warranted. EIA will continue to support such legislation efforts and hopes the Administration will reconsider its opposition in light of the facts presented today.

The EIA Telecommunications Group, including all the major U.S. members, which represent the vast majority of the telephone and radio manufacturers of this country, also has begun a major study of Japanese targeting of our telecommunications equipment industry, similar to the "Houdaille" petition. On the basis of the results of that study, it is quite possible that we will file a formal petition under one or more statutory provisions. I must say we are proceeding with this effort in spite of the discouraging response to date of the Administration to the Houdaille petition.

Frankly, experiences like Houdaille and the TV dumping case clearly demonstrate that existing laws are inadequate to deter foreign targeting practices. Some of us, therefore, are seeking changes in U.S. trade laws that will require the U.S. government to monitor foreign practices, self-initiate investigations and impose remedies against all unfair practices, and assure those remedies effectively compensate U.S. industry for damages, and prevent further damages.

We will also seek legislation that will deny to foreign producers the "Act of State" defense for cartel behavior that affects competition in the U.S. market. Why shouldn't they obey our anti-trust laws in this market? That doctrine should be irrelevant to a situation where a foreign government is an active part of a conspiracy to harm U.S. industries. These types of changes are essential to restoring an environment where U.S. businesses in targeted sectors will be willing to invest at a rate that will maintain their technological and market-based competitive strengths. I would hope the administration would recognize the need for changes and become a constructive contributor to new legislation. Let me note at this point, that the Danforth Reciprocity Bill is far short of what is needed.

There are several actions the Administration could take that would begin to deal with the basic imbalance between foreign economies that target exports, and the U.S. economy which does not target. You can begin by making use of the trade authorities you have on the books in a number of ways:

First: You can act to give unilateral relief to Houdaille under Section 103. The provision to remove the investment tax credit is in the law now, and was in the law when the Japanese temporary measures were passed and implemented.

Second: Based on the SIA semiconductor study, you could pursue a GATT complaint against Japan on semiconductors if dramatic changes in U.S. participation in Japan do not appear very soon. We believe the U.S. government could and should bring similar complaints for all targeted industries.

Third: You could establish a system which would monitor imports of targeted industries for unfair practices and initiate inquiries when surges occur. Widespread home market protection abroad means that dumping into this market is likely on a broad range of products. And in telecommunications and other advanced electronics sectors we know subsidies are widespread as well. The Administration has the authority to impose such monitoring and should do so. Otherwise U.S. firms will have to bear the additional expense, effort and aggravation to chase particular problems. When unfair practices are widespread such a system is a reasonable response.

Fourth: You can suspend GATT tariff bindings with respect to new products. This would require renegotiation of the balance of concessions under Article XXVIII, but the cost should be minimal since there would be little or no trade yet in such products. A prime candidate for such treatment would be cellular radio. No outsiders have been allowed to test radios in the Japanese cellular system, but six Japanese firms have been permitted to participate in a U.S. test system. When cellular systems go into commercial operation in the near future, it is clear that the Japanese will be poised to exploit the opportunity here, while no more than one U.S. firm—Motorola—appears to have a limited crack at the Japanese market. Suspension of tariff bindings will provide significant leverage with Japan even if those duties are not raised. And an increase in duties could be effected if the U.S. industry is threatened.

Fifth: The U.S. government should change its negotiating objectives vis-à-vis Japan. Instead of seeking access through lower barriers, it should ask for results in the form of increased U.S. participation in the Japanese market. This approach would be consistent with the Japanese government's targeting of results for domestic producers and is realistically the only way we can expect any dramatic improvement of U.S. exports given the strong bias against imports in the Japanese business community. The Japanese government should launch an affirmative action program to persuade the Japanese buyer to buy U.S. products. The U.S. should also ask Japan to terminate its targeting practice in advanced electronics, where Japanese firms are large and successful enough to compete without government support.

Sixth: The U.S. government should match any form of foreign government assistance that adversely affects U.S. firms in export markets. This means not only greater guarantee authority for the Export-Import Bank, but also increased resources in whatever other forms are required, including direct credits. The U.S. should also be willing to use political pressure to counter efforts by foreign competitor governments to influence potential buyers.

The President's State of the Union message said, "This Administration is committed to keeping America the technological leader of the world now and into the 21st century." Later he went on to say, "America must be an unrelenting advocate of free trade. As some nations are tempted to turn to protectionism, our strategy cannot be to follow them, but to lead the way toward freer trade." Unfortunately, we cannot have it both ways. If others persist in targeting and protecting home markets and the U.S. does nothing to protect its interest, we will not be the technological leaders for the rest of this century, much less the next century. In telecommunications, microelectronics and computers we may not even last through this decade in that role.

Let's make no mistake about what will really happen if we don't reverse our approach. The trade problem will become more extreme, radical proposals to close our market will come forward, and be enacted. We will move to an era like the late '20s and '30s with galloping protectionism. The Japanese need a big market to support their export program and keep their industry afloat. This is the biggest available market, but the destruction of this market through over-penetration of key industries will finally destroy the very market they so desperately need.

Some propose we brush aside our anti-trust laws, and join the Japanese move to government-planned markets and allow massive collective efforts. They proposed the U.S. reorder our way of life. If we move in this way, we will see creeping alliance that will lead to cooperation, then consortium in our own country as practiced in Europe and Japan. This movement will eliminate our environment for the entrepreneur to develop new ideas and whole new businesses, as we have seen for many decades in our country, in things like the personal computer. Only the big, the strong and those in cooperation will survive in this kind of market, and then the consumer will demand government involvement. I am not sure we would like to see either of these programs as our future way of life in America. We look to this government for leadership to preserve our economic system.

Export Financing and Licensing Considerations in the Telecommunications Industry

BY JOHN N. LEMASTERS
SENIOR VICE PRESIDENT
COMMUNICATIONS SECTOR
HARRIS CORPORATION

I would like to address two major issues—export financing and licensing—and their detrimental effect on the U.S. telecommunication industry's competitiveness in international markets.

In the case of export financing, the U.S. is at a serious disadvantage versus foreign competition. For example, France, England and Japan routinely offer 85% financing compared with EXIM's 65% ceiling. This forces U.S. suppliers to supplement EXIM projects with higher-cost commercial financing, with the result that American loans are non-competitive. Also, the EXIM interest rate, despite the welcomed reduction, remains generally higher than the rates offered by our foreign competitors.

To alleviate this situation, it is recommended that the Administra-

223

tion supports Senator Heinz Bill S.2600. This calls for the removal of EXIM loan authorization levels from the expenditure section of the Federal Budget. It will also increase appropriations for the Direct Loan Program from the proposed level of $6.5 billion in 1984 to an amount sufficient to allow 85% financing.

Further, it is urged that EXIM interest rates be established at the lower of OECD agreement rates or intermediate U.S. Government borrowing rates plus a modest spread for EXIM's overhead expenses. The recently announced "war chest" appropriation must be used where necessary to combat subsidized overseas financing offers.

EXIM must also be encouraged to adopt a leadership role in long-term funding, particularly where sovereign-risk loans are concerned. Currently, EXIM appears reluctant to approve new loans or extend existing commitments for Latin American countries rescheduling their debts, while overseas agencies such as COFACE, MITI/EXIM Japan, and ECGD are encouraging financing in many of these countries. It should be noted that the IMF, World Bank, and U.S. Treasury are also supporting these same countries during these troubled times.

Further, EXIM should permit 10 to 15 percent of local content financing to be included in a total direct loan package. This would bring the U.S. in line with other OECD members' practices and allow American offers to be less complex and more competitive. The ECGD financing for Eastern Airline aircraft, because of the U.K.'s Rolls-Royce engine content, is an example of the extent of other countries' preparedness to support their national industries.

In the area of military equipment, while EXIM is generally prohibited from financing military purchases of non-strategic equipment, other countries such as England, France and Germany aggressively pursue these opportunities. EXIM should provide financing where U.S. equipment is being upgraded or replaced. It should also consider financing in situations where other OECD members are quoting terms or the equipment has a combined military and civilian use.

Another area in which U.S. companies are at a disadvantage is contractor guarantees. the OPIC program of guaranteeing companies against contract repudiation and unfair calling of contractor bonds was for a brief period administered by EXIM Bank but is now returned to OPIC administration. It is recommended that such bond guarantees be removed from the OPIC program entirely and placed under EXIM's political risk coverage.

Since EXIM has the responsibility of evaluating and accepting repayment risks, it is in the best position to evaluate and cover contractor bonds in those situations where such bonds are most at risk, e.g., when a foreign government repudiates political and business ties with the U.S. and a contractor's bond might be used unfairly by that government. The pre-shipment coverage on FCIA insurance policies is an EXIM program of this type which is already being provided.

These are just some of the more important issues on which the Administration can lend direct and positive support to U.S. industry without violating GATT or any other trade agreements with OECD members. Underlying all of these problems is the apparent attitude that prevails of the EXIM and OPIC program administrations that "the U.S. is doing a favor for overseas customers". This is very different from the hard-hitting marketing attitude of foreign governments.

Turning now to licensing agreements, it is recommended that controls on the export of high technology products be reduced and there be a reciprocal tightening of controls on the underlying technologies involved in such products. Experience in the telecommunications industry shows that generally speaking, by the time a U.S. product is developed the industry is available throughout the OECD countries and industries. Too much time and effort is wasted in the U.S. on the so-called control of mature products which are already commercially available throughout the world. Control of technology must take place prior to the commercial introduction of a new product.

The Export Administration Act of 1979 calls for DOD to develop and publish a list of military-critical technologies which should be controlled, and recognizes that the focus of export controls should be on critical technologies rather than end products. The DOD list has been revised several times and has been under review by indus-

225

try groups and U.S. COCOM partners. The security classification system employed by the DOD should be used to control both the application of key technologies to commercial products utilizing these technologies.

Probably the single most damaging U.S. action affecting American exports, particularly to lesser-developed countries, is the practice of applying export restrictions for foreign policy purposes. This must be stopped. It undermines confidence in U.S. suppliers and encourages customers to turn to other countries whose governments do not have the reputation of stepping in to stop contracted shipments of goods or prohibiting future contracts.

This practice is particularly harmful to the U.S. telecommunications industry which has to contend with numerous and highly aggressive foreign competitors. Customers will not buy telecommunications equipment from a country whose government policies can jeopardize continuity of supply of the spare parts and maintenance service critical to the utilization of equipment. There is no doubt that the U.S. telecommunications industry is being severely crippled by this policy.

Finally, it is recommended that the Department of Commerce, and all other government agencies concerned with finding solutions to the U.S. imbalance of trade develop organizations specifically charged with aiding and supporting exports. The U.S. no longer has such a great technological lead that it is a highly sought-after supplier to world markets. In telecommunications, most OECD countries have equal product capability as a result of the technology transfer that has been taking place since World War II.

Despite this highly competitive world arena, U.S. companies have difficulty in finding an organization within any government department that will help with detailed problems of licensing and financing from a common sense, pro-export view. Most departments are tied-up with the control aspect of exporting. It is very difficult for a large company to operate in this environment and almost impossible for a small company to deal with this control bureaucracy.

Why not establish a project office for each major segment of industry, such as telecommunications, and give that office an export goal to achieve over a period of time? These offices should have sufficient technical expertise to understand and monitor foreign competitive product offerings. They should also be charged with helping U.S. companies overcome control barriers and financing restrictions so that American products can be actively and aggressively promoted in international markets.

![black bar]

Final Overview

BY RICHARD E. HORNER
CHAIRMAN AND CHIEF EXECUTIVE OFFICER
E. F. JOHNSON COMPANY

An overview of the industry perspective on trade matters must inevitably take into account the broad diversity of experience and opinions of individual companies and their leaders. However, the steady growth of trade in manufactured products over the past three decades—with the resulting implications for economic health in the United States—has generated an interest level that promotes a common understanding of many of the economic, technological, political and sociological facts that impact trade results. Such a common understanding provides a departure point for the more qualitative and subjective criteria that would bear on any consensus for an action program to improve the performance of the United States in world trade. I will attempt a brief review of the circumstances that condition our consideration of remedial action.

229

The United States experienced a net deficit in its merchandise trade accounts in 1982 in excess of $42 billion even though petroleum imports were down $17 billion, and the export of military hardware was up $15 billion. The deficit with Japan widened by almost $19 billion.

A recent report from the Treasury Department projects a potential deficit in the current year at $75 billion.

Inasmuch as the country had a favorable balance of trade in products from the extractive industries with Japan, the impact on the labor and capital of the manufacturing industry in the United States was even greater than indicated by the net trade balance.

The deficit in current accounts, although much smaller, was negative and disruptive in monetary and fiscal policy, especially when considered in context with international credit transactions.

The 1982 report of the AMEX Bank of London shows the LDCs $600 billion in debt to the financial capacity of developed countries, with the observation that there is no possibility of servicing that debt from resources developed by current levels of trade. A substantial fraction of that debt is held by U.S. commercial banks, and virtually all of it is a threat to world economic health and a drag on international trade.

The U.S. domestic economy is in serious difficulty. After a series of progressively shorter economic cycles, with our political leadership vacillating between the fiscal and monetary cures for inflation and unemployment, we find ourselves emerging from the longest and deepest recession since World War II with current inflation, unemployment, interest and deficit spending rates at intolerable levels. It is a sign of the times that a 5% inflation rate gives the appearance of being under control by comparision with the other parameters of our economy.

The degree to which our trade deficits serve as cause or effect of the performance of the national economy is subject to some argument, but political pragmatism recognizes that the Japanese do not vote in the United States' national elections.

There is general agreement that government deficits, the high cost of borrowed money, and unemployment are tightly interdependent and bear heavily on the competitive capability of our industrial economy.

Our government has been reduced to a selection between two remedies for deficit spending: reduction in government spending and the increase of tax revenues. Reduced spending has increased faster than revenues even though inflation, the imbedded increases in payroll taxes, and the so-called Tax Equity and Fiscal Responsibility Act continue to take larger bites each year out of the gross national product.

In this economic environment, government authorities charged with the responsibility of administering our national trade policy have struggled with a set of trade laws and international agreements inherited from an earlier time when our export commerce was dominated by commodities and the products of our extractive industries. The task of negotiating new relationships responsive to the needs of high technology industry, especially as it is organized and implemented by small entrepreneurial companies that have the best opportunity for multiplying job opportunities, has fallen way short of the target of timely change to match the dynamism of evolving technology and international competition. The entrepreneurial spirit of small companies and the fluidity of equity capital are totally defeated by the years of adjudication required under current trade laws in the interest of due process and what turns out to be the Achilles heel that is successfully attacked by the unrestrained endurance of trading competitors whose human and capital resources are organized and implemented on a national basis.

Japan has used the mechanics of industry and market targeting with a liberal interpretation and in some instances, the violation of the rules of the General Agreement on Tariffs and Trade, to overwhelm our markets. The process of market designation, the closing of Japanese domestic markets to imports, the marshalling of resources by cartelization and subsidy of industry, and the cooperative support of the attack on U.S. markets, with the joint efforts of trading companies, financial institutions and product companies, is generally recognized as the framework for their success. We have seen many variations on this formula in the process of losing major fractions—and sometimes control of—our markets for automotives, steel, machine tools and consumer electronics. Our alarm today is related to their designation of semiconductors, computers, and telecommunications which strikes at the core of our industrial capacity of the future. We have no reason to believe that the practices of the past, together with techniques not yet called upon, will not be successful in overwhelming these markets and crippling our economy if the U.S. does not mount an effective response. The telecommunications sector is particularly vulnerable to unfair trade practices since it is controlled by a regulatory authority in the market of each of our trading partners to the advantage of their domestic industry.

The U.S. Government is seriously handicapped in responding to this challenge unless it recognizes its weaknesses and uses its strengths. Our government and industry organization is based on a maximum degree of freedom and diversity. The three branches of government and the separate regulatory agencies have a high degree of independence and operate as a check and balance on each other. Competition between and within the various branches of government is frequently more immediate and sometimes appears to be more important than competition outside our national boundaries. As an industrial and military leader in the world, we have been able to afford internal adversarial relationships, but we are being attacked through that weakness. We need to recognize that attack and counter with our strengths.

Our political system has naturally promoted adversarial relationships between government and business. Organizations don't vote; people do. Therefore, there is a temptation to make our business organizations responsible for collecting taxes and correcting all of the ills of the society. Improvements in pollution, product safety, workplace safety standards, unemployment compensation and a long list of needs can be satisfied with hidden costs to our society if they are submerged in the accounting of business balance sheets.

It is time that we remembered that our strength lies in our free market economy and our political democracy. It is the operation of our Government's fiscal and monetary policies in the framework of broad ownership of our industrial capacity that made our strength and that now debilitates it. This is the basis for the following suggestions to stop our slide to a position of mediocrity in the world economy at great cost to our standard of living. It is not clear that the world can prosper without our leadership. The remedial actions required may seem to be of cosmic proportions. That is only because they are a match for the magnitude of the problems.

It is the process of formulating an action program to seek trade balance that raises sharp disagreements. The fact that the playing field in the export-import game is tilted toward our markets and away from those of our trading partners is fairly well established by the results, but this is not uniformly hurtful to all Americans at the same time. Consumers and importers rather enjoy buying products at less than fully absorbed costs; and, in the short term, the consequences of unemployment and disruption in the capital markets are scattered. In fact, the longer-term penalty to our standard of living will probably be blamed on some cause that is closer to home. The handiest target to absorb the blame is a host of shortcomings that can be classified under the title of industrial efficiency. Management ineptness, capital shortages, union greed, addiction to quarterly earnings and the rejection of the work ethic are just a few of the ills that must be repaired to restore our competitive position. Indeed, industrial management has much opportunity to improve performance; and perfection is an ever-receding goal. However, I believe there is also universal agreement that the government has a role in guaranteeing a level playing field, and that is the focus of the balance of my comments for which I would not claim universal endorsement.

First, let's recognize the merit and the limitations of government-to-government trade negotiations. The merits are related primarily to long-term continuity of communications and the formulation and understanding of the boundary conditions for trade relationships. They are generally useless and misleading in their efforts to work fundamental changes in the policies of political democracies. Every political leader understands that he cannot serve his country if he is thrown out of office by its electorate. This gives him a focus in time that is explicitly related to the next election which generates an absolute requirement for understanding and agreement on the need for change by the *people* of his country, not just the trade negotiators.

233

Endless negotiations and unfulfilled promises result from setting un-
achievable objectives. Some objectives necessary to our national interest must be attained through our actions, not our trade diplomacy.

A second recommendation is both subjective and substantive.
Let's stop invoking the terrors of trade retaliation and world-wide
depression resulting from the Smoot-Hawley model of tariff barriers. In the first place, the model is fifty years out of date, and the
world of trade that involves the United States has changed in revolu-
tionary ways. We do have the largest monolithic market with the
lowest entry requirements for any comparable opportunity in the
world. We are a world class leader in technology, capital formation,
labor efficiency and management sciences. We fulfill a similar role
in the defense of the free world and in the promotion of human
rights. We frequently do not get the respect that our leadership responsibility requires for reasonable success. Free trade in the world
requires discipline just as the other freedoms do. For some reason,
we seem to understand the importance of preparedness and deterrence in preserving peace and political freedoms, but we are in-
clined toward unilateral disarmament in the arena of world trade. If
we are too scared to assert the authority of the leadership mantle that
is endowed by the wealth of our markets, then, indeed we will lose
them; and their worth will be destroyed in the process.

The next four recommendations have to do with areas of government activity that will require constant attention to repair mistaken
government policies of the past that have so handicapped our trading performance as to make us a sitting duck for the sharpshooting
trade policies of the Japanese, and soon, the Europeans.

The first area is that of our tax policy. Perhaps we could get
philosophical agreement that all of the costs of government and government programs of wealth redistribution are ultimately paid for by
consumption. Taxes on capital are especially self-defeating. They
deplete our store of capital, negate positive motivational forces, and
pervert our political system. Our government reached a new high in
abstract morality last year in labeling the Tax Equity and Fiscal Responsibility Act. The only thing that was equitable about it was the
willingness of Congress to share their Constitutional responsibility
for raising taxes by asking business to increase their collection ac-

tivities. This may sell to the voters, but it is an administrative and fiscal nightmare and a long step backward from the encouragement of capital formation.

Our tax statute is perilously close to a degree of complexity that defies understanding, encourages the underground economy by destroying discipline, and defeats the long-range planning that is essential to a competitive position in the world of trade. Changing its course will require the abandonment of political expediency and a re-education process internal to government, but it must be done.

A close companion to our government's money-raising policies is our government's money-spending policies. As a leader in the world of trade, we do not have the freedom to devalue our currency as though it were a real or a peso. Our deficits go up, and the government inclination is to raise taxes and increase its borrowings. Interest rates go up; inflation goes up; unemployment rises. The result is further increases in the deficit. In lockstep with the rise in interest rates, the market value of the dollar increases and imports are up while exports go down—another contribution to unemployment. All because our government doesn't know how to stop spending. Coming from the Midwest as I do, I can tell you there is an awful lot of government spending out there that is less than essential. The government budgeting process struggles with programs and their constituencies. The support of any program that doesn't increase faster than the inflation rate is described as a draconian cut and such terms are anathema to our political process. I have a suggestion derived from industrial management experience. Every business manager knows there is only one way to reduce spending, and that is to reduce employment. People not only draw salaries and fringe benefits, they develop programs that cost money. There are a few principles of government budget management that can be based solidly on experience. First, tax increases will not reduce deficits—in fact, all of our fiscal history indicates exactly the opposite—and we have arrived at the highest era of government deficits simultaneously with the government tax take not only at its highest point in absolute numbers, but at a record fraction of GNP. Does anybody remember that TEFRA was to produce $100 billion in revenue and $300 billion in savings? Were we destined to have deficits $400 billion higher than those that are now projected without TEFRA? If it were possible in government as it is in industry to arbitrarily reduce employment by 10% at all grade levels, government spending would promptly recede by a comparable fraction. The essential elements of all programs would be accomplished—probably more efficiently than now—but the rate of new initiatives for the use of the government dollars would be reduced, and the health of our economy would respond like magic. This is a model that has been used successfully in industry over and over again, but of course, in industry the managers are elected by the owners. Their responsibility to the employees is primarily to maintain a viable enterprise. The suggested formula would be very difficult in government, but unless our government gets its fiscal house in order, a competitive trade posture is impossible and of secondary importance.

Another area of government responsibility that is directly related to its fiscal policy is the monitoring and modulation of our commercial banking systems. There is a myth about the land that the prime rate is an important benchmark against which capital costs are measured, and it is currently set at the highest real margin over the inflation rate in history because of future uncertainty incident to government borrowing. I have never noticed any reluctance on the part of the market-center banks to raise the prime rate every week if current operations require it. I presume it could be lowered at the same rate. There are certainly adequate instruments available to protect midterm and long-term commitments. After all, bankers still work on the spread between the cost and price of money. The unique problem of the banking systems today is the unusual fraction of assets that are in non-producing and non-collateralized loans. The domestic economy is subsidizing an aggressive loan policy that has been endangered by domestic and worldwide recessions. Our government, in promoting the interests of segments of our export economy and a stable world economy, has encouraged commercial loans that have come home to roost without debt service. Of course, some of the current exposed position can be attributed to the avarice of bank management, but it is a regulated industry. If we are to have domestic interest rates free of the burden of the reserves required for banking the LDCs, then we had better be a little more careful about merging the interests of our commercial banking system with those of the World Bank, the IMF, and the EXIM Bank.

In each of these recommendations, you will recognize in your own reaction a degree of implausibility because the suggested remedies are not consistent with past government practice and political practicality. I would be the first to agree that a major problem of this country is economic illiteracy. We are suffering the growth pangs of a nation that makes its living as a sophisticated industrial society whose people are exposed every day to an information transfer process the substance of which is, at best, confusing. The elected government is understandably trying to respond to the mixed signals from the electorate while simultaneously exercising leadership. How can that leadership be effective—especially related to the issues of world trade—if different segments of government are trying to pull in different directions? Simultaneously, there is a desperate need for public education on issues of trade economics. Some government entity (perhaps the Commerce Department) needs to take the lead in a communications program expressing trade policy and the basic economic issues involved. One of the strengths of the Japanese program is their well-orchestrated effort to communicate what they are doing to the participants in Japan and to their customers in the U.S. If our government doesn't know where it is going in trade policy, it probably doesn't make much difference how we get there; but I suspect there is going to be a national consensus in favor of restoring reasonable balances in the next couple of years. The suggestion that our deficits might double in 1983 over that experienced in 1982 will have a cataclysmic impact on our political system if we don't change the game plan soon.

Finally, there is one action that can be taken explicit to current trade performance that will be more effective in the short term than any other tool that is available. I refer, of course, to Section 103 of the 1971 Revenue Act. I referred earlier to the importance of our leadership role in assuring free trade and the effectiveness of deterrence in establishing discipline to level the playing field for all trade contestants. This provision in our law which permits the President to withhold the investment tax credit for the purchase of any capital equipment imported from a nation that engages in unfair trade policies probably needs to be renewed with an implementation procedure that is believable, responsive, and somewhat protected from the constraints of diplomatic and defense relations. The provision has been in the law for more than ten years. A case has been before the President for many months, and the very delay that it has experienced suggests that there is great reluctance on the part of our government in exercising its essential leadership role. It probably will be necessary to use the provision a few times to test its effectiveness and make it believable as a deterrent. As an instrument of free trade enforcement, it is potentially very efficient. This is especially true in the high technology telecommunications market segment. It can be made to be immediately responsible in tactical situations. It probably is more than 100% efficient in terms of its economic impact; that is to say, by making every industrial buyer in the United States conscious of the decision that he is making in buying capital equipment imports from a country that has been designated as unfair in its trade practices, the disadvantage to the seller will be greater than just the economic impact of the tax credit. The effect will thus extend beyond just the capital equipment markets. It is non-trivial that the government has an opportunity to take immediate action in the approval of the Houdaille Petition. It is hoped that it would be approved with a public statement that announces to all the world that we are serious about free trade; that free trade means balanced trading conditions; and that in each case where we find it necessary to exercise control over our markets, we will be glad to describe in specifics the necessary qualifications for removal of the threat of further deterrence.

Telecommunications, like computers, are provided in this country and sold throughout the world by a large variety of industrial concerns. Gigantic companies (some of the largest in the world) and tiny entrepreneurial companies are equally important in forming the total fabric of our national future in these market segments. We cannot afford a trade system that ignores the needs of any part of this industry. Our present trade rules are not responsive to the dynamics of the industry or the technology. Our political system has discerned this fact. If the Administration doesn't move quickly, we are at risk to some pretty heavy-handed remedies generated by our political process. By working together, we can fulfill our leadership responsibilities to the advantage of free trade in the world and a healthier world economy.

Summary of Specific Issues Raised by the Telecommunications Industry

Summarizing the current issues of particular concern to the telecommunications industry:

■ Performance requirements are perceived as forms of protectionist barriers preventing market access to some foreign countries. Foreign markets are essentially closed by post, telegraph and telephone agencies (PTTs) while the U.S. market is open to all foreign manufacturers.

■ Foreign countries' targeting practices are believed to diminish U.S. manufacturers' market share of both domestic and foreign markets and represents a serious threat to its viability.

■ The industry believes that present export licensing procedures hinder or prevent product export. Industry spokesmen were concerned that export controls not be used as foreign policy instruments. They argued that controls should be reduced on products and that procedures should be developed to protect export of the underlying technology.

■ Discussion of the Nippon Telephone and Telegraph (NTT) agreement demonstrated that a majority of companies thought the agreement has not produced significant results, although one company suggested that the agreement could be a model for similar agreements.

■ Project financing support and interest rates from the Eximbank were considered to be non-competitive.

■ Export promotion support from U.S. embassies was considered to be helpful. However, the industry expressed frustration with the difficulties in identifying specific U.S. Government offices to assist with detailed problems of export licensing, financing and ultimate sales.

Index

243

INDEX